Crop Water Productivity of Winter Wheat at Multiscale and Its Improvements over the Huang-Huai-Hai Plain, China

Edited by Liu Qin, Sarah Garré, Yan Changrong, Zhang Hui, Qu Chunhong

SCIENCE PRESS
Beijing

Crop Water Productivity of Winter Wheat at Multiscale and Its Improvements over the Huang-Huai-Hai Plain, China

Liu Qin, Sarah Garré, Yan Changrong, Zhang Hui, Qu Chunhong

审图号：GS（2018）5446 号
ISBN 978-7-03-055880-0

Copyright© 2019 by Science Press
Published by Science Press
16 Donghuangchenggen North Street
Beijing 100717, China

Responsible Editor: Li Xiuwei

Printed in Beijing

All rights reserved. No part of this publication may be reproduced, stored in a retrieval system, or transmitted in any form or by any means, electronic, mechanical, photocopying, recording or otherwise, without the prior written permission of the copyright owner.

Contributors

Bernard Tychon, Department of Environmental Sciences and Management, Arlon Campus Environment, University of Liège, 185, Avenue de Longwy, 6700 Arlon, Belgium. E-mail: bernard.Tychon@ulg.ac.be

Dong Wenyi, Institute of Environment and Sustainable Development in Agriculture, Chinese Academy of Agricultural Sciences, Beijing 100081, China. E-mail: dongwenyi@caas.cn

Hao Weiping, Chinese Academy of Agricultural Sciences, Beijing 100081, China. E-mail: WeipingHao@caas.cn

He Wenqing, Institute of Environment and Sustainable Development in Agriculture, Chinese Academy of Agricultural Sciences, Beijing 100081, China. E-mail: hewenqing@caas.cn

Huo Zhiguo, Chinese Academy of Meteorological Sciences, Beijing 100081, China. E-mail: huozg@camscma.cn

Jiang Shuai, Huizhou Meteorological Bureau of Guangdong Province, Huizhou 516001, China. E-mail: jiangshuai_2006@126.com

Ju Hui, Institute of Environment and Sustainable Development in Agriculture, Chinese Academy of Agricultural Sciences, Beijing 100081, China. E-mail: juhui@caas.cn

Li Xiangxiang, Center of Agrometeorology of Jiangxi Province, Nanchang 330046, China. E-mail: lixiangxiang0901@163.com

Li Zhen, Institute of Environment and Sustainable Development in Agriculture, Chinese Academy of Agricultural Sciences, Beijing 100081, China. E-mail: lizhen@caas.cn

Liu Canran, Arthur Rylah Institute for Environmental Research, Department of Sustainability and Enviroment, Heidelberg, Victoria 3084, Australia. E-mail: canranliu@hotmail.com

Liu Enke, Institute of Environment and Sustainable Development in Agriculture, Chinese Academy of Agricultural Sciences, Beijing 100081, China. E-mail: liuenke@caas.cn

Liu Qin, Institute of Environment and Sustainable Development in Agriculture, Chinese Academy of Agricultural Sciences, Beijing 100081, China. E-mail: liuqin02@caas.cn

Mei Xurong, Chinese Academy of Agricultural Sciences, Beijing 100081, China. E-mail: meixurong@caas.cn

Qu Chunhong, Agricultural Information Institute, Chinese Academy of Agricultural Sciences, Beijing 100081, China. E-mail: quchunhong@caas.cn

Sarah Garré, University of Liege, Gembloux Agro-BioTech, Department of Biosystems Engineering, Passage des déportés, Gembloux 5030, Belgium. E-mail: sarah.garre@ulg.ac.be

Sun Dongbao, Institute of Environment and Sustainable Development in Agriculture, Chinese Academy of Agricultural Sciences, Beijing 100081, China. E-mail: sundongbao@caas.cn

William D. Batchelor, Biosystems Engineering Department, Auburn University, Auburn, AL, 36849, USA

Xu Jianwen, Dalian Meteorological Service of Liaoning Province, Dalian 116001, China. E-mail: xujianwen0101@163.com

Yan Changrong, Institute of Environment and Sustainable Development in Agriculture, Chinese Academy of Agricultural Sciences, Beijing 100081, China. E-mail: yanchangrong@caas.cn

Yan Zhenxing, Institute of Environment and Sustainable Development in Agriculture, Chinese Academy of Agricultural Sciences, Beijing 100081, China. E-mail: yanzhx@hotmail.com

Yang Jianying, Chinese Academy of Meteorological Sciences, Beijing 100081, China. E-mail: yangjy@camscma.cn

Zhang Hui, Center for Development of Rural Social Undertaking, Ministry of Agriculture and Rural Affairs, Beijing 100122, China. E-mail: zhanghui@agri.gov.cn

Zhao Fenghua, Key Laboratory of Ecosystem Network Observation and Modeling, Institute of Geographic Sciences and Natural Resources Research, Chinese Academy of Sciences, Beijing 100101, China. E-mail: zhaofh@igsnrr.ac.cn

Preface

Drought and water shortage are generally accepted to be one of the most critical problems faced by worldwide agriculture, and it is so especial in China, where agricultural production and prosperity are mainly dependent on the timely, adequate and proper distribution of rainfall. The analysis of water productivity is becoming very critical in light of population growth, food security and increasing pressure on water resources. However, there is limited understanding of the spatio-temporal variation of crop water productivity (CWP) in crop rotation systems and the key factors that influence the Huang-Huai-Hai Plain where groundwater sources were over-exploitation and warmer and drought conditions in the future will intensify crop water demand. As the largest water consumption, the agricultural sector is facing a challenge to produce more crops with less water. Consequently, the Huang-Huai-Hai Plain is facing with the double threats of both maintaining high and stable crop yields and improving CWP of winter wheat through reducing water consumption.

The Huang-Huai-Hai Plain belongs to the extratropical monsoon climatic region. Its annual mean precipitation is 500–600 mm (with more than 70% falls from July to September), while the atmospheric evaporative demand is about 800 mm·y^{-1}. For the wheat-maize rotation system, rainfall can just meet 65% of total agricultural water demand, especially in the growing season of winter wheat, during which only 25%–40% is satisfied by precipitation. Irrigation water is mainly pumped from groundwater sources. Drought is one of the most damaging and widespread extreme climates in the world, and it has been the relative restriction factor in agriculture and economy development in China. As the world's largest water consumption industries, irrigation farming system consumes around 70% (even more than 80% in some countries) of total freshwater usages. However, the accessible freshwater for agriculture has been declining along with the worldwide population boom and increasing water consumption in daily living, industry and environmental protection activities. Additionally, changes in quantitative value and spatial-temporal distribution of solar radiation, heat resources, and precipitation further fluctuate agricultural production. It showed that precipitation had decreased significantly at the amount of 30 mm during the past 50 years in North and Northeast China, inducing a rapidly expanding drought areas. In the background of global warming, the irrigation water resources for agricultural production is expected to be more unstable due to the increasing complexity of the regional difference and the annual/interannual variability of precipitation. Thus, how to produce more food with less water will become a subject to solve urgently in the next decades, and virtually depends on CWP. Thus, improvement of CWP is of primary importance to alleviate water resources crisis, guarantee national food security, and ensure social sustainable development.

In our study, we aimed to identify the most sensitive and primary controlling variables to potential evapotranspiration and in sub-sequence projected drought characteristics under the RCP8.5 scenario, to establish the potential impact of climate change and drought on the winter wheat yield using CERES-Wheat Model in selected locations, and to investigate water productivity of wheat using the SEBAL model, a linear regression equation integrating remotely sensed images and census data, and the grain yield in the Huang-Huai-Hai Plain.

In our work, the major agronomic consequences have been drawn regarding the reform of the common agricultural policy in Huang-Huai-Hai Plain, China. Results and conclusions from the studies presented in this book may be helpful to identify subjects for further research and actions in management to filling the information and knowledge gaps in agricultural water management. Researchers are encouraged to further investigations into how to implement these practices with an emphasis on improving the sustainability of these agro-ecosystems.

This book was edited by the team of scientists mainly based on the Institute of Environment and Sustainable Development in Agriculture, Chinese Academy of Agricultural Sciences and the research platform AgricultureIsLife of the University of Liège-Gembloux Agro-Bio Tech. It was supported by the National Science Foundation of China (41401510, 41675115) and the Agricultural Science and Technology Innovation Program (2017–2020).

Liu Qin
June 30, 2019

Contents

Chapter 1 Climate change and crop water productivity: opportunities for improvement ·· 1
  1.1 Climate change and crop water productivity (CWP) ···························· 2
    1.1.1 Climate change and agricultural production ······························· 2
    1.1.2 The potential evapotranspiration and meteorological drought ······· 3
    1.1.3 Water storage ··· 5
    1.1.4 The impact of climate change on crop yields ····························· 6
    1.1.5 Crop water productivity ··· 7
  1.2 Context, objectives and outline of the book ·· 9
    1.2.1 Context ··· 9
    1.2.2 Objectives ·· 10
    1.2.3 Outline ·· 11
  1.3 Study region and data collection ·· 12
    1.3.1 Study region ··· 12
    1.3.2 Data collection ··· 14
  1.4 Methods ·· 16
    1.4.1 Calculation of potential evapotranspiration ······························ 16
    1.4.2 DSSAT-CERES-Wheat model for yield simulation ··················· 17
    1.4.3 Satellite-based actual evapotranspiration estimation using the SEBAL method ··· 19
  References ··· 20

Chapter 2 Impacts of climate change on potential evapotranspiration under a historical period and future climate scenario in the Huang-Huai-Hai Plain, China ··· 31
  2.1 Introduction ··· 32
  2.2 Materials and methods ·· 34
    2.2.1 Study area ·· 34
    2.2.2 Meteorological data ··· 36
    2.2.3 Estimation of potential evapotranspiration ······························· 37
    2.2.4 Time series analysis to quantify major trends ··························· 37
    2.2.5 Sensitivity analysis and multivariate regression ························ 38

2.3　Results ··· 38
　2.3.1　Historical and future trends of meteorological variables ······················· 38
　2.3.2　Spatial and temporal characteristics of ET_0 ·· 40
　2.3.3　Temporal variation of sensitivity coefficients ······································ 45
　2.3.4　Regional response of ET_0 to climate change ····································· 48
2.4　Discussion ·· 49
　2.4.1　Spatio-temporal evolution of ET_0 ·· 49
　2.4.2　Impact of meteorological variables on ET_0 ······································· 50
　2.4.3　Estimated precipitation deficit and impact on agriculture ···················· 51
2.5　Conclusions ··· 52
References ·· 53

Chapter 3　Spatio-temporal variation of drought characteristics in the Huang-Huai-Hai Plain, China under the climate change scenario ················· 59
3.1　Introduction ·· 60
3.2　Materials and methods ··· 62
　3.2.1　Study region ·· 62
　3.2.2　Climate data ·· 62
　3.2.3　Drought area data ·· 63
　3.2.4　Calculations of drought indices ·· 63
　3.2.5　Drought identification using run theory ··· 64
3.3　Results ··· 65
　3.3.1　Selection of preferable drought index ··· 65
　3.3.2　Drought characteristics over the past 50 years ································· 65
　3.3.3　Drought prediction for 2010–2099 under RCP8.5 scenario ············· 70
3.4　Discussion ·· 73
　3.4.1　Trend variations between different drought indices ························· 73
　3.4.2　Applicability of drought index ··· 75
3.5　Conclusions ··· 76
References ·· 77

Chapter 4　Potential effect of drought on winter wheat yield using DSSAT-CERES-Wheat model over the Huang-Huai-Hai Plain, China ········· 81
4.1　Introduction ·· 81
4.2　Materials and methods ··· 83
　4.2.1　Study region and data description ·· 83
　4.2.2　Calculation of precipitation deficit for winter wheat ······················· 84
　4.2.3　Crop model description ·· 85
　4.2.4　Statistical tests for trend analysis ··· 85
4.3　Results ··· 86

4.3.1	DSSAT evaluation	86
4.3.2	Trends and persistence of typical growth date and precipitation deficit	87
4.3.3	Variation of yield reduction rate	88
4.3.4	Cumulative probability of yield reduction rate	89

4.4 Discussion ········· 91
4.5 Conclusions ········ 92
References ········· 92

Chapter 5 Investigation of the impact of climate change on wheat yield using DSSAT-CERES-Wheat model over the Huang-Huai-Hai Plain, China ········ 97

5.1 Introduction ········ 98
5.2 Materials and methods ········ 100
   5.2.1 Study region ········ 100
   5.2.2 CERES-Wheat crop model ········ 101
   5.2.3 Simulated scenarios: past, future and isolated variables ········ 102
5.3 Results ········ 104
   5.3.1 Testing of CERES-Wheat model ········ 104
   5.3.2 Changes in growth duration and related climate variables ········ 105
   5.3.3 Changes in yield and the contributions of single climate variables ········ 107
5.4 Discussion ········ 109
   5.4.1 Negative impact of increasing solar radiation ········ 109
   5.4.2 Positive impact of warming temperature and increasing precipitation ········ 111
5.5 Conclusions ········ 112
References ········ 113

Chapter 6 The impacts of climate change on wheat yield based on the DSSAT-CERES-Wheat model under the RCP8.5 scenario in the Huang-Huai-Hai Plain, China ········ 116

6.1 Introduction ········ 117
6.2 Materials and methods ········ 118
   6.2.1 Study region ········ 118
   6.2.2 CERES-Wheat model ········ 118
   6.2.3 Simulation design ········ 119
6.3 Results ········ 121
   6.3.1 Model calibration and validation ········ 121
   6.3.2 Simulated changes of the phenological phase ········ 122
   6.3.3 Changes of climatic variables during the wheat-growing period ········ 123
   6.3.4 Impacts of different climate variables on wheat yield ········ 124
   6.3.5 Impact of elevated CO_2 on wheat yield ········ 126

6.4 Discussion ··· 127
　6.4.1 The impact of warming temperatures ··· 127
　6.4.2 Uncertainties ··· 128
6.5 Conclusions ··· 129
References ··· 130

Chapter 7　Water consumption in winter wheat and summer maize cropping system based on SEBAL model in the Huang-Huai-Hai Plain, China ··· 133

7.1 Introduction ··· 134
7.2 Materials and methods ··· 135
　7.2.1 Study area ··· 135
　7.2.2 Crop dominance map ··· 136
　7.2.3 Phenological data ··· 136
　7.2.4 MODIS products ··· 137
　7.2.5 Meteorological data ··· 137
　7.2.6 SEBAL model ··· 137
7.3 Results ··· 140
　7.3.1 Crop ET_a ··· 140
　7.3.2 Correlation among ET_a, NDVI, and land surface temperature ··· 141
　7.3.3 Correlation between ET_a and geographic parameters ··· 143
7.4 Discussion ··· 145
　7.4.1 Assessment of regional crop evapotranspiration ··· 145
　7.4.2 Separation of evapotranspiration of the two crops ··· 146
　7.4.3 Possible uncertainty of results ··· 146
　7.4.4 Need for refinement ··· 147
7.5 Conclusions ··· 147
References ··· 148

Chapter 8　An assessment of water consumption, grain yield and water productivity of winter wheat in agricultural sub-regions of Huang-Huai-Hai Plain, China ··· 152

8.1 Introduction ··· 153
8.2 Materials and methods ··· 154
　8.2.1 Study region description ··· 154
　8.2.2 Data collection ··· 155
　8.2.3 CWP estimation ··· 155
8.3 Results ··· 157
　8.3.1 ET map ··· 157
　8.3.2 Wheat yield map ··· 159

  8.3.3 CWP map ·· 159
  8.3.4 Relations among yield, ET_a and CWP ·· 161
 8.4 Discussion ·· 163
 8.5 Conclusions ··· 165
 References ·· 165

Chapter 9 General discussion, conclusions, and prospects ···································· 170
 9.1 Overview of results and hypotheses ·· 171
  9.1.1 ET_0 and drought characteristics ··· 171
  9.1.2 Effects of climate change and drought on wheat yield ······························ 172
  9.1.3 Spatial variability in crop water productivity ·· 173
 9.2 General discussion ·· 175
  9.2.1 Agricultural adaptations for CWP improvements ······································· 175
  9.2.2 The uncertainties ·· 178
  9.2.3 Referable value from dataset and methodology of this book ···················· 179
 9.3 Conclusions ··· 180
 9.4 Prospects and improvements ··· 181
  9.4.1 Increasing RCP scenarios alternatives ·· 181
  9.4.2 Increasing collection of irrigation and fertilizer management for
     DSSAT simulation ··· 182
  9.4.3 Increasing collection of observed CWP in agro-meteorological stations ········· 183
 9.5 Closing words ··· 183
 References ·· 184

List of abbreviations

3H Plain:	Huang-Huai-Hai Plain
ADAP:	Anthesis days after planting
CMA:	China Meteorological Administration
CWP:	Crop water productivity
DSSAT model:	Decision Support System for Agrotechnology Transfer model
ET_0:	Potential evapotranspiration
ET_a:	Actual evapotranspiration
ET_c:	the water requirement for crops
IPPC:	International Plant Protection Convention
ISODATA:	Iterative self-organizing data analysis technique algorithm
LST:	the land surface temperature
MDAP:	Maturity days after planting
M-K test:	Mann-Kendall test
MODIS:	Moderate-resolution imaging spectroradiometer
NDVI:	Normalized difference vegetation index
P-M:	Penman-Monteith equation
RCP:	Representative concentration pathway
SEBAL:	the surface energy balance algorithm for land model
SPEI:	the standardized precipitation evapotranspiration index

Chapter 1 Climate change and crop water productivity: opportunities for improvement

Abstract

Drought and water shortage are widely accepted to be one of the most critical problems faced by worldwide agriculture, and it is so especial in China where agricultural production and prosperity are largely dependent on the timely, adequate and proper distribution of rainfall. The analysis of water productivity is becoming very critical in light of population growth, food security and increasing pressure on water resources. The Huang-Huai-Hai Plain, located in Northern China, is the major grain producing areas in China. The main crop pattern is winter wheat and summer maize rotation system, which provides about 70% and 30% of wheat and maize production in China, respectively. Due to the extratropical monsoon climate, more than 70% of annual precipitation falls during summer seasons (July to September). Thus, winter wheat growing period has suffered from serious water deficit and only 25%–40% of water demand is satisfied by rainfall. To maintain a high yield, wheat is irritated with pumped groundwater which has led to environmental threats, such as groundwater lowering and land surface subsidence. Additionally, the warming temperature and changing precipitation pattern caused by climate change have induced perturbations to regional crop production. Consequently, the Huang-Huai-Hai Plain is facing with the double threat of both maintaining high and stable crop yields and improving crop water productivity of winter wheat by reducing water consumption. However, there is limited understanding of the spatio-temporal variation of crop water productivity in the rotation system and the key factors that influence the Huang-Huai-Hai Plain. With the projected temperature increase and change in precipitation distribution, the drought risk is expected to increase further and subsequently make crop production more uncertain through augmenting crop water consumption in the study area. According to the predictions of climate change models, ET_0 is expected to increase over the coming years due to an expected temperature rise. The general recognition is that drought has been intensifying around the world due to global warming over the past decades. In this book, we aimed (1) to identify the most sensitive and primary controlling variables to potential evapotranspiration and in sub-sequence projected drought characteristics under the RCP8.5 scenario, (2) to establish the potential impact of climate change and drought on the winter wheat yield using CERES-Wheat model in selected locations, and (3) to investigate water productivity of wheat using the SEBAL model, a linear regression equation which integrates remotely sensed images and census data, and the grain yield in the Huang-Huai-Hai Plain.

1.1   Climate change and crop water productivity (CWP)

1.1.1   Climate change and agricultural production

A global change in the main meteorological variables is observed in the last decades. According to the IPCC report, the global temperature has risen by 0.74°C in recent 100 years (1906–2005), and it is likely to continue in the 21st century, causing changes in the hydrological cycle by affecting precipitation and evaporation (IPCC, 2013). By the end of this century, the global mean temperature could be 1.8–4.0°C warmer than at the end of the previous century (Solomon et al., 2007). Warming will be uneven across the globe and is likely to be greater in lands than in oceans, the poles as well as arid regions (Solomon et al., 2007). Recent weather records also show that land surface temperatures are increasing more slowly than expected from climate models, potentially because of a higher level of absorption of CO_2 by deep oceans (Balmaseda et al., 2013). Similar to the global trend. China also experienced warming trend. The average annual temperature rose by 0.5–0.8°C during the 20th century, and most of the temperature increase took place over the past 50 years (Wu et al., 2016). Furthermore, dry areas in Northern China have been warming faster than wet areas in Southern China (Ding et al., 2007; Piao et al., 2010). At the global scale, precipitation tended to increase in the high latitude regions of the Northern Hemisphere and in the tropical regions; while in the semi-tropical regions, the precipitation decreased over the past several decades (IPCC, 2013). In China, the decrease in annual precipitation was significant in most of Northern China and the eastern part of Northwest China (Bai et al., 2007).

Although there are many impacts expected from global climate change, one of the largest impacts is expected to be on agriculture (Pearce et al., 1996; Piao et al., 2010; Wang et al., 2013b). Climate change, especially the increasing risk of extreme weather events and indirect impacts on freshwater resources, threatens agricultural systems and food security (Barros et al., 2014; Field, 2012; Hertel et al., 2010; Wheeler and von Braun, 2013). Climate change will have important implications in regions with limited water supplies, such as the Huang-Huai-Hai Plain of China, where expected warmer and drier conditions will augment crop water demand. The Huang-Huai-Hai Plain, one of the largest plains in China, is located in Northern China. Water shortage in this region has become a serious concern in recent decades (Brown and Rosenberg, 1997). Excessive exploitation of groundwater resources has resulted in a water table decline at a rate of 1 m·y^{-1} and severe groundwater depression over the past 20 years (Zhang et al., 2015). An expected reduction of annual precipitation and aggravated drought risk will bring negative effect to agricultural production and food security in coming decades over the Huang-Huai-Hai Plain. A more detailed review of climate change impacts on drought and subsequent crop yields is presented in the following sections.

1.1.2 The potential evapotranspiration and meteorological drought

Based on observed and modeling data, numerous studies have demonstrated an increase in the frequency and intensity of droughts (Dai, 2013), which implies a growing threat to food security. Drought conditions will be aggravated due to climate change by increasing potential evapotranspiration and augmenting crop water consumption in water-limited regions (Goyal, 2004; Maracchi et al., 2005; Thomas, 2008).

Potential evapotranspiration (ET_0) is defined as the amount of water that can potentially evaporate and be transpired from a vegetated surface with no restrictions other than the atmospheric demand (Lu et al., 2005), and widely acknowledged as a key hydrological variable representing an important water loss from catchments. According to the predictions of climate change models, ET_0 is expected to increase over the coming years due to an expected temperature rise (Goyal, 2004; McNulty et al., 1997). However, decreasing trends of ET_0 have been detected in some regions of China (Chen et al., 2006; Thomas, 2000; Wang et al., 2007), the United States (Irmak et al., 2012), and Australia (Roderick and Farquhar, 2004). Therefore, global atmospheric temperatures rise will not necessarily give rise to ET_0 in all cases. For example, a reduction in solar radiation can compensate for the impact of temperatures on ET_0 as observed in many places (Stanhill and Cohen, 2001; Wild, 2014). Conventionally, aridity is usually expressed as a generalized function of precipitation, temperature, and ET_0 reflecting the degree of meteorological drought. Over the last decades, the aridity index (which is defined as the ratio of ET_0 to precipitation) displays different trends in different regions. Su et al. (2015) found that drought index had a bigger decreasing trend (at a rate of -0.313 y^{-1}) in winter than other seasons in 1961–2012 according to 11 meteorological stations in and around the Shiyang River Basin. Liu et al. (2013) also reported that annual aridity index decreased significantly by 0.048 y^{-1} and was primarily governed by increasing precipitation according to 80 national meteorological stations in Northwest China from 1960 to 2010.

Given that China, with merely 7%–8% of the world's arable land resources, has to feed 22% of the global population, food security in China is an urgent issue in the context of climate change (Wang et al., 2016). In China, droughts have become more frequent and intense during the last decades, and presented a direct threat to crop growth in vast areas across the country (Dalin et al., 2015; Piao et al., 2010). Over the past six decades, severe droughts hit China in the 1960s, the late 1970s and early 1980s, and the late 1990s (Guan et al., 2015; Zhai and Zou, 2005). Accordingly, there is an urgent demand for effective monitoring of drought stress, especially in areas with limited water resources. Several techniques have been developed to analyze drought characteristics quantitatively (Heim, 2002). Some of these are the physically based indices, such as Palmer drought severity index (PDSI) along with its derivatives, or statistically based indices, such as standardized precipitation index (SPI) and the standardized precipitation evapotranspiration index (SPEI)(Vicente-Serrano et al., 2011). These indices have been widely used in detecting long-term drought trends under climate change at several locations around the world. The general recognition is that drought severity and duration have been intensifying around the world due to global warming over the past decades (Allen et al., 2010; Dai, 2013). However, it has been difficult to understand how droughts have changed in China, because the findings based on the potential evapotranspiration equations vary among studies (Sheffield et al., 2012; Trenberth et al., 2014; Xu et al., 2015). Significant drying trends were detected in Northern and Southwestern China during the past decades, when ET_0 was estimated by temperature only (Wang et al., 2015a; Yu et al., 2014a). However, when the aridity index was calculated by the Penman-Monteith equation with more climatic variables taken into account, no evidence of an increase in drought severity could be found across China (Wang et al., 2015b), and there were even more wetting areas than drying areas observed in Northeast China Plain (Xu et al., 2015).

As a major crop production area in China, the Huang-Huai-Hai Plain has experienced serious drought and water scarcity problems in recent years (Yong et al., 2013), which has been the limiting factor in agricultural production (Zhang et al., 2015). Furthermore, water limitations are likely to be accentuated by increased food demand, soil quality deterioration and over-exploitation of groundwater resources (Yang et al., 2015). Climate variability, especially extreme climate events such as drought, may cause fluctuation of crop yields (Lu and Fan, 2013; Yu et al., 2014b). Thus, understanding the potential variations of drought characteristics under climate change is essential for reducing vulnerability and establishing drought adaptation strategies for agriculture in the Huang-Huai-Hai Plain. Most previous studies have primarily reported the seasonal and spatial variability of water deficiency (Huang et al., 2014; Yong et al., 2013) and the long term drought evolutions, including drying trends, spatial distribution of drought frequency, drought-affected areas, and drought duration for historical periods (Wang et al., 2015a, 2015b; Xu et al., 2015; Yu et al., 2014a). However, few studies have evaluated the performance of multi-indices (such as SPEI-PM, SPEI-TH, and SPI) on estimating drought impact and assessed drought risk for future climate scenarios.

1.1.3 Water storage

Terrestrial water storage is a key component of the global hydrological cycle and plays a critical role in Earth's climate system (Mo et al., 2016). Natural and anthropogenic stresses such as climate change, drought, increasing water use and agricultural practices will affect both surface water and groundwater resources globally (Wang et al., 2013a). The GRACE (the Gravity Recovery and Climate Experiment) satellite mission has proved to be an invaluable tool in monitoring such hydrological changes with global coverage and sufficient spatial and temporal resolution (Ramillien et al., 2008; Rodell et al., 2009) and revealing trends in present-day continental water storage in many parts of the world. Chen et al. (2009) combined GRACE data, hydrological models, and local in situ precipitation data to study annual variations in terrestrial water storage in the Amazon basin and noted that GRACE was capable of observing the extreme Amazon drought event of 2005. In addition, on the basis of the monthly GRACE gravity field, Leblanc et al. (2009) observed changes in surface water storage caused by severe droughts in South Australia between 2003 and 2006, and Feng et al. (2012) used GRACE data to monitor variations in terrestrial water storage in the Amazon basin from 2002 to 2010.

Mo et al. (2016) applied GRACE Tellus products in combination with simulations by Global Land Data Assimilation System (GLDAS) and data from reports, to analyze variations in terrestrial water storage (TWS) in China as a whole and eight of its basins from 2003 to 2013. As described in the figure reported by Mo et al. (2016), from 2003 to 2013, the southwest rivers region and the Huang-Huai-Hai Plain showed significant water storage depletions, and the area of depletion was largest in spring and summer, respectively. Human activities and climate parameters should be responsible for the significant water storage depletion in China, while in the Huang-Huai-Hai Plain, agricultural irrigation consumes large amounts of groundwater pumped from deep wells (Kendy et al., 2004). Usually, the farmers in the Huang-Huai-Hai Plain apply 3–5 times irrigation, sometimes even 6–7 times, through flood irrigation during the winter wheat growing season to get higher grain yields (Li et al., 2010).

1.1.4 The impact of climate change on crop yields

Drought is one of the most damaging and widespread climate events, negatively affecting agricultural production, water resources, ecosystem function and human living around the world (Dai, 2011; Dilley et al., 2005; Wilhite et al., 2007). The simulations by Rosenzweig and Parry (1994) showed that there is a large degree of spatial variation in crop yields across the globe. Both the sign and the magnitude of the projected changes in crop yields varied with alternative climate models and from one country to another. In general, crop yields increased by 30% in Northern Europe, but decreased by around 20% across Asia, Africa and South America between 2050 and 2010 (Rosenzweig and Parry, 1994). As the first example of the global impacts of climate change on crop production, these simulations are remarkable.

Crop production is affected by climatic variables such as rising temperatures, changing precipitation regimes and increased atmospheric CO_2 levels, and also affected by biological variables such as the lengths of the crop growth periods. A longer life cycle was one of the most widely observed biological changes in the response of crops to climatic warming across the Northern Hemisphere during the twentieth century (Steltzer and Post, 2009). Overall, the yields of wheat and maize have responded negatively to the recent warming since the 1980s on a global scale, although the yield response signals of other crops (e.g., rice) are still unclear (Lobell and Field, 2007). Extensive reviews of the drivers of climate change impacts on crop yields are provided by Gornall et al. (2010) and Rosenzweig et al. (2013). As described in Table 1-1, direct drivers of climate change impacts on crop yields include the long-term change in average temperature and precipitation conditions, and the increasing occurrence of extreme weather events such as extreme temperatures, droughts, floods and tropical storms. In addition, crop yields are sensitive to the indirect effects of climate change on freshwater resources, pests and diseases, and sea level rise.

Table 1-1 Direct and indirect drivers of climate change impacts on crop yields according to the report by numerous researchers

Driver	Impact
Mean climate change	Shift in crop growing season (Kucharik, 2008).
Extreme temperature	Heat stress; reduced crop fertility (Semenov and Shewry, 2011).
Drought	Water stress; crop development alteration (Li et al., 2009; Savage, 2013).
Flood	Fungal disease; crop failure (Rosenzweig et al., 2002; Schiermeier, 2011).
Tropical storm	Loss of cropland area; crop failure (Schiermeier, 2011).
Pests and disease	Crop failure; reduced quality (Rosenzweig et al., 2001).
CO_2 increase	Enhanced the photosynthesis rate; reduced stomata transpiration; reduced protein content (Kimball, 2010; Myers et al., 2014; O'leary et al., 2015).

In recognizing the complex interactions between crop growth and environmental factors, numerical simulations, or crop models, have become popular research tools for researchers in agro-meteorology in recent years. Modeling techniques are involved in empirical models (Lobell and Burke, 2010), process-based crop models for detailed biophysical process (Jones et al., 2003; Keating et al., 2003; Parry et al., 2004; Stöckle et al., 2003) and large-scale ecosystem models aiming to simulate the terrestrial carbon cycle (Deryng et al., 2011; Osborne et al., 2007). Bai et al. (2016) investigated that using the Agricultural Production System Simulator (APSIM) model increase in temperature reduced wheat yield by 0.0%–5.8% and decrease in solar radiation reduced it by 1.5%–8.7% in Jiangsu and Anhui provinces of China. Jabeen et al. (2017) also observed that the rise in maximum and minimum temperatures decreased the wheat yields using DSSAT and GIS across the Pothwar region, Pakistan. Deryng et al. (2011) found that changes in temperature and precipitation as predicted by PEGASUS 1.0 (Predicting Ecosystem Goods and Services Using Scenarios) for the 2050s led to a global yield reduction if planting and harvesting dates remain unchanged. Wilcox and Makowski (2014) documented a meta-analysis of simulated wheat yield from 90 studies to identify the levels of temperature, precipitation and CO_2 concentration that resulted in increasing or decreasing wheat yields.

Of these models, the CERES (Crop Estimation through Resource and Environment Synthesis) cereal model simulated the growth and development of cereal crops including wheat in response to weather and management factors (Ritchie and Otter, 1985). The successful performance of CERES-Wheat in simulating wheat growth and grain yield in response to management and environmental factors has been reported under a wide range of soil and climatic conditions (Attia et al., 2016; Feng et al., 2016; Jabeen et al., 2017; Thorp et al., 2010; Timsina et al., 2008; Zheng et al., 2016). Hence, DSSAT-CERES has been used for climate change and climate extreme impact assessment for rice, wheat and maize for different zones in China in historical and future scenarios (Feng et al., 2016; Liu et al., 2012; Xiong et al., 2009; Zheng et al., 2016).

1.1.5 Crop water productivity

Crop water productivity (CWP) is defined as the amount of yield produced per unit of water involved in the production, or the value added to water in a given circumstance (Molden et al., 1998; Sakthivadivel et al., 1999; Tuong et al., 2000), which is a quantitative term used to define the relationship between crop produced and the amount of water involved in crop production and consequently becomes a useful indicator to quantify the impact of irrigation scheduling decisions with regard to water management (Igbadun et al., 2006). The threat of climate change is affecting the important water resources (Chatterjee et al., 2012), and subsequent precipitation variability and changing evaporation rate will lead to variations in water availability and groundwater recharge (Huntington, 2006). The increased frequency of extreme weather events such as drought disaster, extreme high temperature, and heavy precipitation have started creating imbalances in the hydrological cycle and resulted in large fluctuations in crop yields and water productivity in recent years.

The Huang-Huai-Hai Plain faces a serious threat of excessive exploitation of groundwater resources (Jia and Liu, 2002; Liu et al., 2010). The potential impacts of climate change are expected to reshape the patterns of demand and supply of water for agriculture. Therefore the assessment of the impacts of climate change on ecological and agricultural water consumption is essential. Understanding the quantity of agricultural water consumption is a high priority in areas where water is currently scarce and over-exploited (Perry, 2011). Evapotranspiration (ET) is a useful indicator of crop water consumption; therefore, accurate estimation of regional ET is essential to achieving large scale water resources management (Rwasoka et al., 2011). Current estimates of actual evapotranspiration in China are mainly based on plot-scale experiments (Chen et al., 2002; Zhang et al., 1999, 2004), from the product of soil moisture and potential ET. However, such estimates are only useful for a specific area, and cannot be expanded to large-scale areas. The level of water consumption differs significantly across regions, farming systems, canal command areas, and farms (Molden et al., 2003).

Development of remote sensing technology has made it possible to estimate land surface evapotranspiration at regional or basin scale. Numerous remote sensing methods for modeling actual evapotranspiration (ET_a) of crop have been improved in recent years (de Oliveira et al., 2009; Jia et al., 2012; Teixeira et al., 2009). Li et al. (2008) estimated the ET_a for winter wheat in Hebei province in the North China Plain (NCP) using the SEBAL (Surface Energy Balance Algorithm for Land) model and NOAA (National Oceanic and Atmospheric Administration) data. Liaqat et al. (2015) also investigated the ET_a for the Indus Basin Irrigation System in Pakistan using SEBS (the Surface Energy Balance System) and MODIS (Moderate-resolution imaging spectroradiometer) products. However, observed phenological data and crop dominance map were not considered in these investigations, and specific water consumption of winter wheat and summer maize has not yet been determined in the Huang-Huai-Hai Plain.

Crop water productivity is a useful indicator for quantifying the impact of irrigation management decisions (Ali and Talukder, 2008) and can be used to assess and compare the effects of water-saving measures at different scales and under various conditions (Cui et al., 2007). Perry et al. (2009) reported in their review that the relationship between biomass and transpiration is essentially linear for a given crop and climate- provided nutrients are adequate. Other results, however, suggest the relationship between ET and yield is not always linear. Most of the above studies reporting on the relationship between water consumption, grain yield, and CWP obtained over a relatively short period, are based on data obtained in a controlled, experimental environment.

1.2 Context, objectives and outline of the book

1.2.1 Context

As described before, a better understanding of the impact of climate change on different weather variables and consequently ET_0 and grain yield is uttermost important to come to a sustainable agricultural production in the Huang-Huai-Hai Plain in China, which is the wheat production base of the country (Figure 1-1). The annual rainfall concentrates in the summer period, from July to September. However, winter and spring are strongly characterized by a lack of water for agricultural production (Yang et al., 2013a). Water shortage in this region has become a serious concern in recent decades (Liu et al., 2010). Excessive exploitation of groundwater resources has resulted in water fallen at a rate of 1 $m \cdot y^{-1}$ and severe groundwater depression in the past 20 years. Furthermore, the yield damage caused by drought tends may increase in the future, indicated by a pronounced uprising of drought events in terms of magnitude and area (Li et al., 2015). As a result, questions related to the implementation of soil conservation and alternative tillage techniques and their impact on the winter wheat yield and evapotranspiration are strongly oriented to developing more sustainable agriculture.

Nevertheless, the extent to which the grain yield and crop water productivity for winter wheat respond to climate change and drought together with their improvement measures across our agricultural region remains unexplored.

Numerous studies have investigated the effect of climatic drought on the grain yield and water productivity and used multi-indices to evaluate drought characteristics, but only for historical climate scenarios (Allen et al., 2010; Dai, 2013). Crop models have been used to assess the impact of climate change and climatic extremes on rice, wheat and maize production for the whole growing season in China (Feng et al., 2016; Liu et al., 2012; Xiong et al., 2009; Zheng et al., 2016). Furthermore, most studies addressing the relationship between water consumption, grain yield, and CWP used data obtained over a relatively short period, or based on data obtained in a controlled, experimental environment (Perry et al., 2009), which leaves us with questions about the variability on the long term.

Therefore, the impact of climate change and drought on simulated yield and land surface evapotranspiration estimation in the Huang-Huai-Hai Plain is of primary interest in the context of improving the crop water productivity for winter wheat while minimizing water consumption in this region. This research can contribute to the establishment of a policy to realize more efficient use of water resources and sustainable agricultural production in the Huang-Huai-Hai Plain, China.

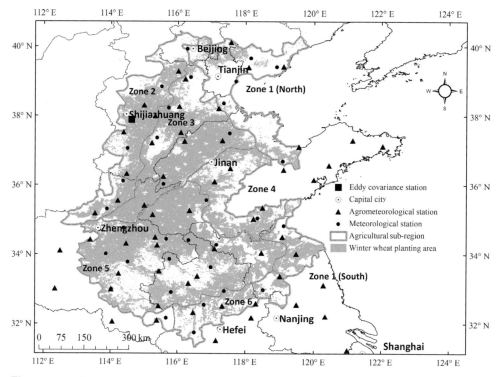

Figure 1-1 The crop dominance map extraction of winter wheat (the green shape) in the Huang-Huai-Hai Plain

1.2.2 Objectives

In this book, we aim (1) to identify the most sensitive and primary controlling variables to potential evapotranspiration and in sub-sequence projected drought characteristics under the RCP8.5 scenario, (2) to establish the potential impact of climate change and drought on the winter wheat yield using CERES-Wheat model over 12 selected locations, and (3) to investigate water productivity of wheat using the SEBAL model, a linear regression equation which integrates remotely sensed images and census data, and the grain yield in the Huang-Huai-Hai Plain.

To achieve these goals, we attempt to answer the following scientific questions in the context of the Huang-Huai-Hai Plain.

(1) What are the characteristics of ET and drought in various climate scenarios and how are they related to water consumption?

(2) How is winter wheat yield affected by climate change and drought?

(3) What are the average water productivity of winter wheat and its spatial variability in the Huang-Huai-Hai Plain?

With the projected temperature increase and change in precipitation distribution, the drought risk is expected to increase further and consequently make crop production more uncertain through augmenting crop water consumption in the study area. According to the predictions of climate change models, ET_0 is expected to increase over the coming years due to an expected temperature rise (Goyal, 2004; McNulty et al., 1997). The general recognition is that drought has been intensifying around the world due to global warming in the past decades (Allen et al., 2010; Dai, 2013). Dynamic crop models, such as the Erosion Productivity Impact Calculator (EPIC) model (Williams et al., 1983), APSIM (Keating et al., 2003), and DSSAT-CERES (Jones et al., 1998), have been tested and used in quantifying responding to water, nitrogen and weather at scales ranging from fields to regions around the world. Jabeen et al. (2017) found that the rise in maximum and minimum temperatures decreased the wheat yields using DSSAT and GIS across the Pothwar region of Pakistan.

Besides the development of remote sensing technology, there have been a number of reports on the crop water productivity in China, with varied spatial and temporal resolutions based on model simulation, estimation from remote sensing data and calculation from collected irrigation district data (Cao et al., 2015; Huang and Li, 2010; Liu et al., 2007; Rosegrant et al., 2002; Yan and Wu, 2014).

Climate change is widely accepted as one of the most critical problems faced by the Huang-Huai-Hai Plain, an groundwater over-exploitation region. Consequently, to answer aforemetioned scientific questions, we would test the following hypotheses (Table 1-2).

Table 1-2 Statement of our hypotheses in this research

Hypothesis	Statement
Hypothesis 1	Drought conditions will aggravate due to climate change by increasing potential evapotranspiration and augmenting crop water consumption in water-limited Huang-Huai-Hai Plain.
Hypothesis 2	The response of winter wheat yield to climate change and drought can be identified available throughout the plain.
Hypothesis 3	The relationship between water productivity with ET and grain yield of winter wheat should be defined at the level of the sub-agricultural zone in the Huang-Huai-Hai Plain.

1.2.3 Outline

This book is a compilation of scientific manuscripts. It is structured as follows.
- Chapter 2 (Hypothesis 1) investigates the spatio-temporal patterns of ET_0 and primary driving meteorological variables based on a historical period and RCP8.5

scenario daily data set from 40 weather stations over the Huang-Huai-Hai Plain using linear regression, spline interpolation method, partial derivative analysis and multivariate regression.
- Chapter 3 (Hypothesis 1) addresses the variations in drought characteristics (drought event frequency, duration, severity, and intensity) for the past 50 years (1961–2010) and under future scenarios (2010–2099), based on the observed meteorological data and the RCP8.5 projection.
- Chapter 4 (Hypothesis 2) determines the potential impacts of drought on wheat yield using the CERES-Wheat model at twelve stations representing different locations in the Huang-Huai-Hai Plain. The cumulative probability of the yield reduction rate during the jointing-heading stage and the filling stage will be investigated.
- Chapter 5 (Hypothesis 2) describes the effects of climate change (including solar radiation, maximum temperature, minimum temperature, and precipitation) based on historical records (1985–2014) and future climate projections under the RCP4.5 and RCP8.5 pathways (2021–2050) on winter wheat yield, using the CERES-Wheat model.
- Chapter 6 (Hypothesis 2) determines the relative impact of each variable shift (temperature, solar radiation, precipitation and elevated CO_2) on wheat yield separately during baseline period (1981–2010), short-term period (2010–2039), medium-term period (2040–2069) and long-term period (2070–2099) under the RCP4.5 and RCP8.5.
- Chapter 7 (Hypothesis 3) documents an attempt to apply a regional evapotranspiration model (SEBAL) and crop information for assessment of regional crop (winter wheat and summer maize) actual evapotranspiration (ET_a) in the Huang-Huai-Hai Plain.
- Chapter 8 (Hypothesis 3) describes spatio-temporal characteristics of water productivity of winter wheat in the Huang-Huai-Hai Plain and in subsequent correlation of water productivity to actual evapotranspiration and grain yield of winter wheat in agricultural subregion over the Huang-Huai-Hai Plain, China.

Finally, in Chapter 9 we discuss the main results and also conclude the book with prospects and potential improvements.

1.3 Study region and data collection

1.3.1 Study region

Climate change is one of the most critical problems faced by the Huang-Huai-Hai Plain, where groundwater is over-exploited, and future warmer and drought conditions will intensify crop water demand. The Huang-Huai-Hai Plain covers the middle and lower reaches of the Yellow River Basin, the Huaihe River Plain, and the Haihe River Plain, extending over 32°00′–40°30′N and 113°00′E to the east coast (Figure 1-1). It is surrounded by the south foot of Yanshan Mountains to the north, north foot of Tongbai Mountains and Dabie Mountains and Jianghuai Watershed to the south, and eastern foot of Taihang Mountain and Qinling Range to the west, whereas the eastern boundary lies west of the Bohai Sea and the Yellow Sea.

The Huang-Huai-Hai Plain belongs to the extratropical monsoon climatic region. The annual mean precipitation is 348.5 mm, while the atmospheric evaporative demand is about 1000 mm·y^{-1}(Ren et al., 2008). As described in Figure 1-2, annual precipitation is concentrated in summer (July through September) and winter is strongly characterized by a lack of water for agricultural production. Water shortages in the Huang-Huai-Hai Plain have become of considerable concern over recent decades (Liu et al., 2010). Increasing trends of 0.003, 0.018 and 0.034°C·y^{-1}, were detected for maximum temperature (T_{max}), average temperature (T_a) and minimum temperature (T_{min}) in the past 54 years, respectively, and it was statistically significant ($P < 0.01$) only for T_a and T_{min} (Figure 1-3). The main cropping system is a rotation of wheat and maize with the systematic application of irrigation water and fertilizer in the Huang-Huai-Hai Plain (Liang et al., 2011; Sun et al., 2011; Zhao et al., 2006), and rainfall can meet just 65% of total agricultural water demand for this rotation system, especially in the case of winter wheat, for which only 25%–40% of the demand is satisfied with rainfall. (Mei et al., 2013b). More frequent irrigation is necessary for high yield levels under these climatic conditions. The irrigation water is primarily pumped from groundwater. Usually, the farmers in the Huang-Huai-Hai Plain apply

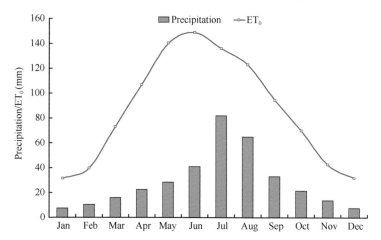

Figure 1-2 Comparison of monthly precipitation and ET$_0$ in the past 54 years. The monthly area-averaged precipitation and ET$_0$ were based on date obtained from 40 meteorological stations across the Huang-Huai-Hai Plain

3–5 times irrigation, sometimes even 6–7 times, through flood irrigation during the winter wheat growing season (Li et al., 2010); this may result in a potential low irrigation water use efficiency and inefficient use of nitrogen (Wang et al., 2004). However, the groundwater level has decreased from a depth of 10 m in the 1970s to 32 m in 2001, and has continued to decrease at the rate of 1 m per year (Zhang et al., 2015).

The plain encompasses around 18 million hectares farmland of which about 61% and 31% are dedicated to wheat and maize production respectively (Jin et al., 2009), with intensive management characterized by the application of sufficient irrigation water and fertilizers. Drought is one of the most damaging and widespread climatic event challenging the Huang-Huai-Hai Plain, and consequently the main restriction factor in agriculture and economy development. Winter wheat is sown in early October and harvested in June of the second year, and that summer maize is then sown immediately afterward and harvested later in September. The Huang-Huai-Hai Plain is divided into six agricultural sub-regions: coastal land, a farming-fishing area (including the northern part, Zone 1, and the southern part), the piedmont plain-irrigable land (Zone 2), the low plain-hydropenia irrigable land and dry land (Zone 3), the hill-irrigable land and dry land (Zone 4), the basin-irrigable land and dry land (Zone 5) and the hill-wet hot paddy field (Zone 6) in terms of climate conditions and agricultural management practices.

1.3.2 Data collection

As mentioned in our study, the first aim of the book is to identify the most sensitive and primary controlling variables of ET_0 and in subsequent projected drought characteristics under the alternative period or scenario in Chapter 2 and 3. A historical dataset from 1961 to 2014, composed of data from 40 meteorological stations, was provided by the China Meteorological Administration. Daily maximum temperature (T_{max}, °C), average temperature (T_a, °C) and minimum temperatures (T_{min}, °C), average relative humidity (RH, %), wind speed (WS, m·s^{-1}) observed at a 10 m height, and daily sunshine duration (SD, h) were available.

The 0.5° × 0.5° gridded data of the Huang-Huai-Hai Plain from 2015 to 2099 were simulated under the future RCP8.5 climatic scenario and provided by the National Climate Center, included daily average temperature (T_a, °C), daily maximum (T_{max}, °C) and minimum temperature (T_{min}, °C), daily precipitation (mm), daily average wind speed (WS, m·s^{-1}), daily average relative humidity (RH, %), and daily net radiation (RS, MJ·m^{-2}·d^{-1}). The 0.5° × 0.5° gridded data were upscaled to 40 meteorological stations using double-linear interpolation method (Yuan et al., 2012). The RCP8.5 scenario is characterized by a high concentration of greenhouse gas with stabilizing emissions post-2099 (increase by about 120 Gt CO_2-eq. by 2100 compared to 2000) (Riahi et al., 2011).

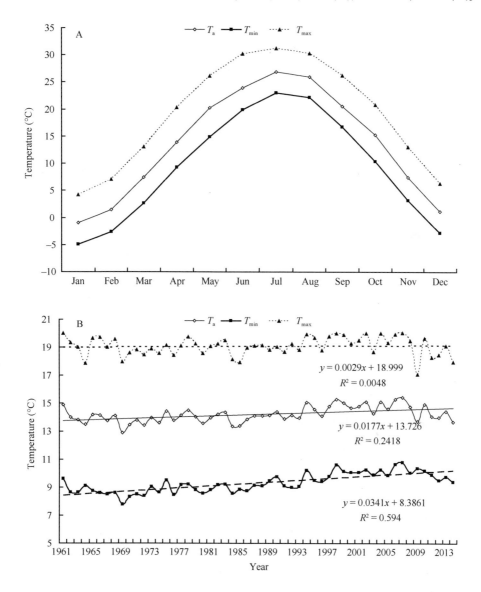

Figure 1-3 Comparison of monthly (A) and yearly (B) maximum (T_{max}), minimum (T_{min}) and average (T_a) temperature in the past 54 years. The monthly and yearly area-averaged three temperatures were obtained from 40 meteorological stations across the Huang-Huai-Hai Plain

Among these 40 stations, we selected 12 stations with more detailed information available. The minimal data sets required for model operation include weather information (daily global solar radiation, maximum and minimum temperatures, precipitation), soil information (classification and basic profile characteristics by soil layer) and management information (e.g., cultivar, planting, irrigation and fertilization information). In this study, the soil classification and profile characteristics were collected from the China Soil Scientific Database.

Finally, we also used satellite data to estimate water productivity in the framework of this project. The data are mainly involved in MODIS products including MOD11A1 (land surface temperature/surface emissivity), MOD13A2 (NDVI) and MCD43B3 (surface albedo). The spatial resolution of the three MODIS products is 1 km. The temporal resolution of MOD11A1, MOD13A2 and MCD43B3 was 1 d, 16 d, and 8 d, respectively. For land surface temperature images, cloudy areas were replaced by the average of two images in the nearest clear dates.

1.4 Methods

1.4.1 Calculation of potential evapotranspiration

Potential evapotranspiration is an integrated climate variable that gives a measure of the evaporation demand of the air. ET_0 is essentially dependent on four meteorological variables: air temperature, solar radiation, relative humidity and wind speed (Allen et al., 1998). The main advantage of the P-M approach is that it takes into account the most significant variables, such as temperature, relative humidity, solar radiation and wind speed as required by physically based equations (Allen et al., 1998), and the influence of each can be analyzed (Blaney, 1952; Mei et al., 2013a). It is also recommended as the standard ET_0 method, by which the evapotranspiration of a hypothetical reference vegetated field is explicitly determined (Allen et al., 2006; Cai et al., 2007). Estimation of the evolution of ET_0 and meteorological variables affecting ET_0 is necessary (Mínguez et al., 2007), and past trends of meteorological variables and ET_0 values still need to establish a solid baseline for future adaptation strategies (Mearns et al., 2003).

We used the Penman–Monteith formula recommended by the Food and Agriculture Organization (FAO) of the United Nations to calculate the ET_0 for both historical and future conditions:

$$ET_0 = \frac{0.408\Delta(R_n - G) + \gamma \dfrac{900}{T+273} U_2(e_s - e_a)}{\Delta + \gamma(1 + 0.34 U_2)} \qquad \text{Formula 1-1}$$

where, ET_0 is potential evapotranspiration (mm·d^{-1}); Δ represents the slope of the saturation vapor pressure/temperature curve (kPa·°C^{-1}); G is the soil heat flux (MJ·m^{-2}·d^{-1}); T is the average daily temperature at 2 m height (°C); U_2 is the wind speed at 2 m height (m·s^{-1}); e_s is the saturation vapor pressure (kPa); e_a is the actual water vapor pressure (kPa); e_s–e_a is the vapor pressure deficit (kPa); γ is the psychometric constant (kPa·°C^{-1}); R_n is the net radiation from the canopy (MJ·m^{-2}·d^{-1}), which is the difference between net shortwave radiation (R_{ns}) and net longwave radiation (R_{nl}). R_{ns} is calculated as

$$R_{ns} = (1-\lambda)R_s \quad \text{Formula 1-2}$$

where, R_s is surface solar radiation (MJ·m^{-2}·d^{-1}) and λ (=0.23) is the albedo of reference grassland. R_s can be calculated from sunshine duration:

$$R_s = \left(a_s + b_s \frac{n}{N}\right)R_a \quad \text{Formula 1-3}$$

where, N is the maximum possible sunshine duration (h), n/N is the relative sunshine duration, R_a is the extraterrestrial radiation (MJ·m^{-2}·d^{-1}), and a_s and b_s are regression constants and set as 0.25 and 0.5 respectively. R_a is calculated as

$$R_a = \frac{24(60)}{\pi} G_{sc} dr \left[w_s \sin(\phi)\sin\xi + \cos(\phi)\cos\xi \sin w_s\right] \quad \text{Formula 1-4}$$

where, G_{sc} is the solar constant (=0.082 MJ·m^{-2}·d^{-1}), dr is inverse relative distance Earth-Sun, w_s is the sunset hour angle (rad), ϕ is the latitude (rad), ξ is the solar declination (rad). R_{nl} is calculated as

$$R_{nl} = \sigma\left[\frac{T_{max,K}^4 + T_{min,K}^4}{2}\right]\left(0.34 - 0.14\sqrt{e_a}\right)\left(1.35\frac{R_s}{R_{so}} - 0.35\right) \quad \text{Formula 1-5}$$

where, σ is the Stefan-Boltzmann constant (=4.903×10^{-9} MJ·K^4·m^{-2}·d^{-1}), $T_{max,K}$ is the maximum absolute temperature during the 24 h period, $T_{min,K}$ is the minimum absolute temperature during the 24 h period, e_a is the actual vapor pressure (kPa), and R_{so} is the clear-sky radiation (MJ·m^{-2}·d^{-1}).

1.4.2 DSSAT-CERES-Wheat model for yield simulation

Crop models can be used to analyze the effects of various climatic factors on crop growth and grain yield considering the interactions with edaphic, biotic and agronomic factors. One of them, DSSAT is an integrated software comprising different computer programs that can simulate crop growth and grain yield for research and decision-making. The latest version DSSAT 4.6.1.0 (http://dssat.net/downloads/dssat-v46) includes Cropping System Model (CSM), the primary modules of which consisted by weather module, soil module, soil-plant-atmosphere module, management operation module, and 27 individual plant growth modules (Figure 1-4). Each plant growth is a separate module to simulate phenology, biomass, growth and yield, based on the soil water and fertility supplement in response to weather and management (Jones et al., 2003).

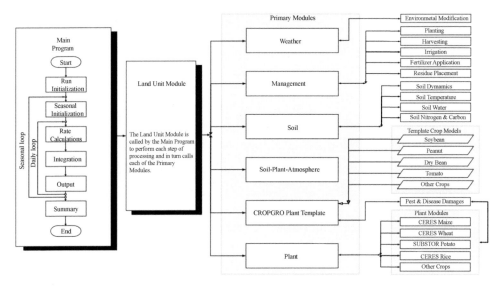

Figure 1-4 Overview of the components and modular structure of the DSSAT/CSM. The DSSAT-CSM has a main drive program, a land unit module, and modules for the primary components that make up a land unit in a cropping system. Each module has six operational steps, as shown in this figure (run initialization, season initialization, rate calculations, integration, daily output, and summary output). The application driver (main program) communicates with only one module of the Land Unit module which provides the interface between the application driver and all of the components that interact in an uniform area of land. Source: Jones et al., 2003

The minimal data sets required for DSSAT model operation consist of weather information (daily global solar radiation, maximum and minimum temperatures, precipitation)(Panda et al., 2003), soil information (classification and basic profile characteristics by soil layer) and management information (e.g., cultivar, planting, irrigation and fertilization information). In the CERES-Wheat model, seven coefficients control the development and growth of wheat (Ritchie et al., 1998), which must be calibrated and evaluated to meet the observed development and growth process under specific environmental conditions before being used for climate impact analyses (Hunt and Boote, 1998).

Due to its capacity of changing the irrigation schedule to simulate plausible climatic drought effects on the crop, the DSSAT CERES-wheat model (He et al., 2013, 2012) was adopted here to simulate the wheat yield during 1981–2014 at all 12 locations. Hence, much work has been done on the impacts of climate change, irrigation scheduling and water and fertilizer conditions on the crop yields using DSSAT in China (Jiang et al., 2016; Ju et al., 2005; Yang et al., 2013b, 2010).

1.4.3 Satellite-based actual evapotranspiration estimation using the SEBAL method

In this study, the Surface Energy Balance Algorithm for Land (SEBAL) method from the meteorological stations and remote sensing data were used to compute the evapotranspiration (ET) rate in the Huang-Huai-Hai Plain in China. The SEBAL method has been used in various studies to assess ET rates in European countries, Southern Asia and China (Bastiaanssen et al., 1998a, 1998b; Wang et al., 1995; Zhang et al., 2016). In addition, Bastiaanssen et al. (2002) have compared predictions of ET and sensible heat flux (H) by SEBAL with measurements made by eddy covariance and scintillometer systems to determine the confidence interval. Comparisons with fluxes measured by other methods confirm the robustness of the SEBAL procedure. A conceptual scheme of SEBAL is presented in Figure 1-5. Hence, various applications have demonstrated the ability of SEBAL to estimate daily ET accurately. The SEBAL algorithm can be applied with little or no ground-based weather data (Allen et al., 2001). When data such as actual measurements for solar radiation and wind speed on the day of the image are available, the predictions by the procedure will be improved. The regional ET was not always available in the resolution of Landsat TM (Thematic Mapper), and SEBAL should be the best way to fix it up.

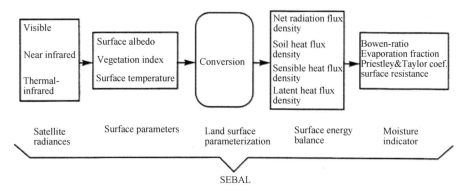

Figure 1-5 Principal components of the Surface Energy Balance Algorithm for Land (SEBAL) which converts remotely measured spectrally emitted and reflected radiance's into the surface energy balance and land wetness indicators (Bastiaanssen et al., 1998a)

The SEBAL procedure consists of a suite of algorithms in this case implemented with the Model Maker module ERDAS software. The algorithms solve the complete energy balance equation:

$$R_n = G + H + \lambda ET \qquad \text{Formula 1-6}$$

where, R_n is the net radiation (W·m^{-2}), G is the soil heat flux (W·m^{-2}), H is the sensible heat flux (W·m^{-2}), and λET is the latent heat flux associated with evapotranspiration (W·m^{-2}).

The net radiation flux on the land surface, R_n (W·m^{-2}), was calculated using the following equation:

$$R_n = (1-\alpha)K_{in} + (L_{in} - L_{out}) - (1-\varepsilon)L_{in} \qquad \text{Formula 1-7}$$

where, α is the surface albedo, K_{in} is the incoming short wave radiation (W·m^{-2}), L_{in} is the incoming long wave radiation (W·m^{-2}), L_{out} is the outgoing long wave radiation (W·m^{-2}), and ε is the land surface emissivity.

The soil heat flux (G) is known to depend on land surface characteristics and soil water content primarily. The soil heat flux was calculated for the SEBAL model by the following equation:

$$G = \frac{T-273.16}{\alpha}\left[0.0032 \times \frac{\alpha}{0.9} + 0.0062 \times \left(\frac{\alpha}{0.9}\right)^2\right](1-0.98NDVI^4)R_n \qquad \text{Formula 1-8}$$

The sensible heat flux (H) was calculated using the following equation:

$$H = \frac{\rho_{air} C_p dT}{r_{ah}} \qquad \text{Formula 1-9}$$

where, ρ_{air} is the air density (kg·m^{-3}), C_p is the air specific heat at constant pressure (J·kg^{-1}·K^{-1}), and r_{ah} is the aerodynamic impedance (s·m^{-1}).

Since the evaporative fraction Λ is constant during a day, the daily ET$_{24}$ (mm) can be estimated using the following equations:

$$\Lambda = \frac{\lambda ET}{R_n - G} \qquad \text{Formula 1-10}$$

$$ET_{24} = \frac{\Lambda(R_{24} - G_{24})}{\lambda} \qquad \text{Formula 1-11}$$

where, ET$_{24}$ is the daily instantaneous evapotranspiration (mm), R_{24} is the daily net radiation (W·m^{-2}), G_{24} is the daily soil heat flux (W·m^{-2}), and λ is the latent heat of vaporization (MJ·kg^{-1}).

References

Ali M H, Talukder M S U. 2008. Increasing water productivity in crop production: A synthesis.

Agricultural Water Management, 95(11): 1201-1213.

Allen C D, Macalady A K, Chenchouni H, Bachelet D, McDowell N, Vennetier M, Kitzberger T, Rigling A, Breshears D D, Hogg E T, Gonzalez P. 2010. A global overview of drought and heat-induced tree mortality reveals emerging climate change risks for forests. Forest Ecology and Management, 259(4): 660-684.

Allen R G, Morse A, Tasumi M, Bastiaanssen W, Kramber W, Anderson H. 2001. Evapotranspiration from Landsat (SEBAL) for water rights management and compliance with multi-state water compacts. IEEE International Geoscience & Remote Sensing Symposium.

Allen R G, Pereira L S, Raes D, Smith M. 1998. Crop evapotranspiration-Guidelines for computing crop water requirements. FAO Irrigation and Drainage Paper 56. Food and Agriculture Organization of the United Nations, Rome, Italy.

Allen R G, Pruitt W O, Wright J L, Howell T A, Ventura F, Snyder R, Itenfisu D, Steduto P, Berengena J, Yrisarry J B, Smith M. 2006. A recommendation on standardized surface resistance for hourly calculation of reference ETo by the FAO56 Penman-Monteith method. Agricultural Water Management, 81(1): 1-22.

Attia A, Rajan N, Xue Q, Nair S, Ibrahim A, Hays D. 2016. Application of DSSAT-CERES-Wheat model to simulate winter wheat response to irrigation management in the Texas High Plains. Agricultural Water Management, 165: 50-60.

Bai A, Zhai P, Liu X. 2007. Climatology and trends of wet spells in China. Theoretical and Applied Climatology, 88(3-4): 139-148.

Bai H, Tao F, Xiao D, Liu F, Zhang H. 2016. Attribution of yield change for rice-wheat rotation system in China to climate change, cultivars and agronomic management in the past three decades. Climatic Change, 135(3-4): 539-553.

Balmaseda M A, Trenberth K E, Källén E. 2013. Distinctive climate signals in reanalysis of global ocean heat content. Geophysical Research Letters, 40(9): 1754-1759.

Barros V R, Field C B, Dokke D J, Mastrandrea M D, Mach K J, Bilir T E, Chatterjee M, Ebi K L, Estrada Y O, Genova R C, Girma B. 2014. Climate change 2014: impacts, adaptation, and vulnerability. Part B: regional aspects. Contribution of Working Group II to the fifth assessment report of the Intergovernmental Panel on Climate Change.

Bastiaanssen W G, Ahmad M D, Chemin Y. 2002. Satellite surveillance of evaporative depletion across the Indus Basin. Water Resources Research, 38(12): 9-1-9-9.

Bastiaanssen W G, Pelgrum H, Wang J, Ma Y, Moreno J F, Roerink G J, Van der Wal T. 1998b. A remote sensing surface energy balance algorithm for land (SEBAL). Part 2: Validation. Journal of Hydrology, 212(1-4): 213-229.

Bastiaanssen W, Menenti M, Feddes R, Holtslag A. 1998a. A remote sensing surface energy balance algorithm for land (SEBAL). Part 1. Formulation. Journal of Hydrology, 212: 198-212.

Blaney H F. 1952. Determining water requirements in irrigated areas from climatological and irrigation data. USDA Soil Conserv. Serv. SCS-TP96: 44.

Brown R A, Rosenberg N J. 1997. Sensitivity of crop yield and water use to change in a range of climatic factors and CO_2 concentrations: a simulation study applying EPIC to the central USA. Agricultural and Forest Meteorology, 83(3-4): 171-203.

Cai J, Liu Y, Lei T, Pereira L S. 2007. Estimating reference evapotranspiration with the FAO Penman–Monteith equation using daily weather forecast messages. Agricultural and Forest

Meteorology, 145(1-2): 22-35.

Cao X, Wang Y, Wu P, Zhao X. 2015. Water productivity evaluation for grain crops in irrigated regions of China. Ecological Indicators, 55: 107-117.

Chatterjee S K, Banerjee S, Bose M. 2012. Climate Change Impact on Crop Water Requirement in Ganga River Basin, West Bengal, India. Third International Conference on Biology, Environment and Chemistry IPCBEE. Singapore: IACSIT Press: 17-20.

Chen J H, Kan C E, Tan C H, Shih S F. 2002. Use of spectral information for wetland evapotranspiration assessment. Agricultural Water Management, 55(3): 239-248.

Chen J, Wilson C, Tapley B, Yang Z, Niu G. 2009. 2005 drought event in the Amazon River basin as measured by GRACE and estimated by climate models. Journal of Geophysical Research: Solid Earth, 114: B05404. doi: 10.1029/2008JB006056.

Chen S, Liu Y, Thomas A. 2006. Climatic change on the Tibetan Plateau: potential evapotranspiration trends from 1961–2000. Climatic Change, 76(3-4): 291-319.

Cui Y, Dong B, Li Y, Cai X. 2007. Assessment indicators and scales of water saving in agricultural irrigation. Transactions of the Chinese Society of Agricultural Engineering, 23(7): 1-7.

Dai A. 2011. Drought under global warming: a review. Wiley Interdisciplinary Reviews Climate Change, 2(1): 45-65.

Dai A. 2013. Increasing drought under global warming in observations and models. Nature Climate Change, 3(1): 52-58.

Dalin C, Qiu H, Hanasaki N, Mauzerall D L, Rodriguez-Iturbe I. 2015. Balancing water resource conservation and food security in China. Proceedings of the National Academy of Sciences, 112(15): 4588-4593.

de Oliveira A S, Trezza R, Holzapfel E A, Lorite I, Paz V P S. 2009. Irrigation water management in Latin America. Chilean Journal of Agricultural Research, 69(Suppl 1): 7-16.

Deryng D, Sacks W, Barford C, Ramankutty N. 2011. Simulating the effects of climate and agricultural management practices on global crop yield. Global Biogeochemical Cycles, 25(2): 2-19.

Dilley M, Chen R S, Deichmann U, Lerner Lam A L, Arnold M. 2005. Natural disaster hotspots: a global risk analysis. Washington: World Bank Publications: 1-145.

Ding Y, Ren G, Zhao Z, Xu Y, Luo Y, Li Q, Zhang J. 2007. Detection, causes and projection of climate change over China: an overview of recent progress. Advances in Atmospheric Sciences, 24(6): 954-971.

ECSNCCA. 2011. Second National Climate Change Assessment Report. Beijing: Science Press.

Feng H, Li Z, He P, Jin X, Yang G, Yu H, Yang F. 2016. Simulation of Winter Wheat Phenology in Beijing Area with DSSAT-CERES Model, Computer and Computing Technologies in Agriculture IX: 9th IFIP WG 5.14 International Conference, CCTA 2015, Beijing, China.

Feng W, Lemoine J M, Zhong M, Tou-Tse H. 2012. Terrestrial water storage changes in the Amazon basin measured by GRACE during 2002–2010. Chinese Journal of Geophysics, 55(3): 814-821.

Field C B. 2012. Managing the Risks of Extreme Events and Disasters to Advance Climate Change: Adaptation: Special Report of the Intergovernmental Panel on Climate Change. Cambridge. Cambridge University Press.

Gornall J, Betts R, Burke E, Clark R, Camp J, Willett K, Wiltshire A. 2010. Implications of climate change for agricultural productivity in the early twenty-first century. Philosophical Transactions of the Royal Society of London B: Biological Sciences, 365(1554): 2973-2989.

Goyal R. 2004. Sensitivity of evapotranspiration to global warming: a case study of arid zone of Rajasthan (India). Agricultural Water Management, 69(1): 1-11.

Guan Y, Zheng F, Zhang P, Qin C. 2015. Spatial and temporal changes of meteorological disasters in China during 1950–2013. Natural Hazards, 75(3): 2607-2623.

He J, Cai H, Bai J. 2013. Irrigation scheduling based on CERES-Wheat model for spring wheat production in the Minqin Oasis in Northwest China. Agricultural Water Management, 128(10): 19-31.

He J, Dukes M D, Hochmuth G J, Jones J W, Graham W D. 2012. Identifying irrigation and nitrogen best management practices for sweet corn production on sandy soils using CERES-Maize model. Agricultural Water Management, 109(9): 61-70.

Heim R R. 2002. A Review of twentieth-century drought indices used in the United States. Bulletin of the American Meteorological Society, 83(8): 1149-1165.

Hertel T W, Burke M B, Lobell D B. 2010. The poverty implications of climate-induced crop yield changes by 2030. Global Environmental Change, 20(4): 577-585.

Huang F, Li B. 2010. Assessing grain crop water productivity of China using a hydro-model-coupled-statistics approach. Part II: Application in breadbasket basins of China. Agricultural Water Management, 97(9): 1259-1268.

Huang Y et al. 2014. Surface water deficiency zoning of China based on surface water deficit index (SWDI). Water Resources, 41(4): 372-378.

Hunt L A, Boote K J. 1998. Data for model operation, calibration, and evaluation. *In*: Tsuji G Y, Hoogenboom G, Thornton P K. Understanding Options for Agricultural Production. Dordrecht: Springer Netherlands: 9-39.

Huntington T G. 2006. Evidence for intensification of the global water cycle: review and synthesis. Journal of Hydrology, 319(1): 83-95.

Igbadun H E, Mahoo H F, Tarimo A K, Salim B A. 2006. Crop water productivity of an irrigated maize crop in Mkoji sub-catchment of the Great Ruaha River Basin, Tanzania. Agricultural Water Management, 85(1): 141-150.

IPCC. 2013. The Physical Science Basis Contribution of Working Group I to the Fifth Assessment Report of the Intergovernmental Panel on Climate Change. USA: Cambridge University Press.

Irmak S, Kabenge I, Skaggs K E, Mutiibwa D. 2012. Trend and magnitude of changes in climate variables and reference evapotranspiration over 116-yr period in the Platte River Basin, central Nebraska-USA. Journal of Hydrology, s 420-421(4): 228-244.

Jabeen M, Gabriel H F, Ahmed M, Mahboob M A, Iqbal J. 2017. Studying Impact of Climate Change on Wheat Yield by Using DSSAT and GIS: A Case Study of Pothwar Region, Quantification of Climate Variability, Adaptation and Mitigation for Agricultural Sustainability. New York: Springer International Publishing: 387-411.

Jia J S, Liu C M. 2002. Groundwater dynamic drift and response to different exploitation in the North China Plain: A case study of Luancheng County, Hebei Province. Acta Geographic Sinica (Chinese edition), 57(2): 201-209.

Jia Z, Liu S, Xu Z, Chen Y, Zhu M. 2012. Validation of remotely sensed evapotranspiration over the Hai River Basin, China. Journal of Geophysical Research: Atmospheres, 117: D13113. doi: 10.1029/2011JD017037.

Jiang Y, Zhang L, Zhang B, He C, Jin X, Bai X. 2016. Modeling irrigation management for water conservation by DSSAT-maize model in arid northwestern China. Agricultural Water Management, 177: 37-45.

Jin H, Wang Q J, Li H W, Liu L J, Gao H W. 2009. Effect of alternative tillage and residue cover on yield and water use efficiency in annual double cropping system in North China Plain. Soil and Tillage Research, 104(1): 198-205.

Jones J W, Hoogenboom G, Porter C H, Boote K J, Batchelor W D, Hunt L A, Wilkens P W, Singh U, Gijsman A J, Ritchie J T. 2003. The DSSAT cropping system model. European Journal of Agronomy, 18(3): 235-265.

Jones J W, Tsuji G Y, Hoogenboom G, Hunt L A, Thornton P K, Wilkens P W, Imamura D T, Bowen W T, Singh U. 1998. Decision support system for agrotechnology transfer: DSSAT v3. *In*: Tsuji G Y, Hoogenboom G, Thornton P K. Understanding Options for Agricultural Production. Netherlands, Dordrecht: Springer: 157-177.

Ju H, Xiong W, Xu Y L, Lin E D. 2005. Impacts of Climate Change on Wheat Yield in China. Acta Agronomica Sinica, 31(10): 1340-1343.

Keating B A, Carberry P S, Hammer G L, Probert M E, Robertson M J, Holzworth D, Huth N I, Hargreaves J N, Meinke H, Hochman Z, McLean G. 2003. An overview of APSIM, a model designed for farming systems simulation. European Journal of Agronomy, 18(3): 267-288.

Kendy E, Zhang Y, Liu C, Wang J, Steenhuis T. 2004. Groundwater recharge from irrigated cropland in the North China Plain: case study of Luancheng County, Hebei Province, 1949–2000. Hydrological Processes, 18(12): 2289-2302.

Kimball B A. 2010. Lessons from FACE: CO_2 effects and interactions with water, nitrogen, and temperature. *In*: Hilell D, Rosenzweig C. Handbook of Climate Change and Agroecosystems: Impacts, Adaptation, and Mitigation. London, UK: Imperial College Press: 87-107.

Kucharik C J. 2008. Contribution of planting date trends to increased maize yields in the central United States. Agronomy Journal, 100(2): 328-336.

Leblanc M J, Tregoning P, Ramillien G, Tweed S O, Fakes A. 2009. Basin-scale, integrated observations of the early 21st century multiyear drought in southeast Australia. Water Resources Research, 45(4): 546-550.

Li H, Zheng L, Lei Y, Li C, Liu Z, Zhang S. 2008. Estimation of water consumption and crop water productivity of winter wheat in North China Plain using remote sensing technology. Agricultural Water Management, 95(11): 1271-1278.

Li Q, Dong B, Qiao Y, Liu M, Zhang J. 2010. Root growth, available soil water, and water-use efficiency of winter wheat under different irrigation regimes applied at different growth stages in North China. Agricultural Water Management, 97(10): 1676-1682.

Li Y, Huang H, Ju H, Lin E, Xiong W, Han X, Wang H, Peng Z, Wang Y, Xu J, Cao Y. 2015. Assessing vulnerability and adaptive capacity to potential drought for winter-wheat under the RCP8.5 scenario in the Huang-Huai-Hai Plain. Agriculture, Ecosystems & Environment, 209: 125-131.

Li Y, Ye W, Wang M, Yan X. 2009. Climate change and drought: a risk assessment of crop-yield impacts. Climate Research, 39(1): 31-46.

Liang W L, Carberry P, Wang G Y, Lü R H, Lü H Z, Xia A P. 2011. Quantifying the yield gap in wheat-maize cropping systems of the Hebei Plain, China. Field Crops Research, 124(2):

180-185.

Liaqat U W, Choi M, Awan U K. 2015. Spatio-temporal distribution of actual evapotranspiration in the Indus Basin Irrigation System. Hydrological Processes, 29(11): 2613-2627.

Liu H L, Yang J Y, Ping H, Bai Y L, Jin J Y, Drury C F, Zhu Y P, Yang X M, Li W J, Xie J G, Yang J M. 2012. Optimizing parameters of CSM-CERES-Maize model to improve simulation performance of maize growth and nitrogen uptake in northeast China. Journal of Integrative Agriculture, 11(11): 1898-1913.

Liu J, Wiberg D, Zehnder A J, Yang H. 2007. Modeling the role of irrigation in winter wheat yield, crop water productivity, and production in China. Irrigation Science, 26(1): 21-33.

Liu S, Mo X, Lin Z, Xu Y, Ji J, Wen G, Richey J. 2010. Crop yield responses to climate change in the Huang-Huai-Hai Plain of China. Agricultural Water Management, 97(8): 1195-1209.

Liu X, Zhang D, Luo Y, Liu C. 2013. Spatial and temporal changes in aridity index in northwest China: 1960 to 2010. Theoretical and Applied Climatology, 112(1): 307-316.

Lobell D B, Burke M B. 2010. On the use of statistical models to predict crop yield responses to climate change. Agricultural and Forest Meteorology, 150(11): 1443-1452.

Lobell D B, Field C B. 2007. Global scale climate-crop yield relationships and the impacts of recent warming. Environmental Research Letters, 2(1): 014002.

Lu C, Fan L. 2013. Winter wheat yield potentials and yield gaps in the North China Plain. Field Crops Research, 143: 98-105.

Lu J, Sun G, Mcnulty S G, Amatya D M. 2005. A Comparison of six potential evapotranspiration methods for regional use in the Southwestern United States. JAWRA Journal of the American Water Resources Association, 41(3): 621-633.

Maracchi G, Sirotenko O, Bindi M. 2005. Impacts of present and future climate variability on agriculture and forestry in the temperate regions: Europe. Climatic Change, 70(1-2): 117-135.

McNulty S G, Vose J M, Swank W T. 1997. Regional hydrologic response of loblolly pine and air temperature and precipitation. Journal of the American Water Resources Association, 33(5): 1011-1022.

Mearns L, Giorgi F, McDaniel L, Shields C. 2003. Climate Scenarios for the Southeastern US Based on GCM and Regional Model Simulations, Issues in the Impacts of Climate Variability and Change on Agriculture. New York: Springer International Publishing: 7-35.

Mei X R, Kang S Z, Yu Q, Huang Y F, Zhong X L, Gong D Z, Huo Z L, Liu E K. 2013b. Pathways to synchronously improving crop productivity and field water use efficiency in the North China Plain. Scientia Agricultura Sinica, 46(6): 1149-1157.

Mei X R, Zhong X L, Vincent V, Liu X Y. 2013a. Improving water use efficiency of wheat crop varieties in the North China Plain: review and analysis. Journal of Integrative Agriculture, 12(7): 1243-1250.

Mendelsohn R. 2010. World Development Report 2010: Development and Climate Change World Bank. Journal of Economic Literature, 48(48): 786-788.

Mínguez M I, Ruiz-Ramos M, Díaz-Ambrona C H, Quemada M, Sau F. 2007. First-order impacts on winter and summer crops assessed with various high-resolution climate models in the Iberian Peninsula. Climatic Change, 81(sup.): 343-355.

Mo X, Wu J, Wang Q, Zhou H. 2016. Variations in water storage in China over recent decades from GRACE observations and GLDAS. Natural Hazards and Earth System Sciences, 16(2): 469-482.

Molden D J, Murray-Rust H, Sakthivadivel R, Makin I W. 2003. A water-productivity framework for understanding and action. *In*: Kijne J W, Barker R, Molden D J. Water Productivity in Agriculture: Limits and Opportunities for Improvement. Sri Lanka, Colombo: CABI Publishing: 332.

Molden D J, Sakthivadivel R, Perry C J, De Fraiture C. 1998. Indicators for comparing performance of irrigated agricultural systems, 20. Srilanka, Colombo: IWMI Research Report.

Myers S S, Zanobetti A, Kloog I, Huybers P, Leakey A D, Bloom A J, Carlisle E, Dietterich L H, Fitzgerald G, Hasegawa T, Holbrook N M. 2014. Increasing CO_2 threatens human nutrition. Nature, 510(7503): 139-142.

O'leary G J, Christy B, Nuttall J, Huth N, Cammarano D, Stöckle C, Basso B, Shcherbak I, Fitzgerald G, Luo Q, Farre-Codina I. 2015. Response of wheat growth, grain yield and water use to elevated CO_2 under a free-air CO_2 enrichment (FACE) experiment and modelling in a semi-arid environment. Global Change Biology, 21(7): 2670-2686.

Osborne T M, Lawrence D M, Challinor A J, Slingo J M, Wheeler T R. 2007. Development and assessment of a coupled crop-climate model. Global Change Biology, 13(1): 169-183.

Panda R, Behera S, Kashyap P. 2003. Effective management of irrigation water for wheat under stressed conditions. Agricultural Water Management, 63(1): 37-56.

Parry M L, Rosenzweig C, Iglesias A, Livermore M, Fischer G. 2004. Effects of climate change on global food production under SRES emissions and socio-economic scenarios. Global Environmental Change, 14(1): 53-67.

Pearce D W, Cline W R, Achanta A N, Fankhauser S, Pachauri R K, Tol R S, Vellinga P. 1996. The social costs of climate change: greenhouse damage and the benefits of control. Climate change 1995. Economic and Social Dimensions of Climate Change: 179-224.

Perry C, Steduto P, Allen R G, Burt C M. 2009. Increasing productivity in irrigated agriculture: agronomic constraints and hydrological realities. Agricultural Water Management, 96(11): 1517-1524.

Perry C. 2011. Accounting for water use: Terminology and implications for saving water and increasing production. Agricultural Water Management, 98(12): 1840-1846.

Piao S, Ciais P, Huang Y, Shen Z, Peng S, Li J, Zhou L, Liu H, Ma Y, Ding Y, Friedlingstein P. 2010. The impacts of climate change on water resources and agriculture in China. Nature, 467(7311): 43-51.

Ramillien G, Famiglietti J S, Wahr J. 2008. Detection of continental hydrology and glaciology signals from GRACE: a review. Surveys in Geophysics, 29(4-5): 361-374.

Ren J, Chen Z, Zhou Q, Tang H. 2008. Regional yield estimation for winter wheat with MODIS-NDVI data in Shandong, China. International Journal of Applied Earth Observation and Geoinformation, 10(4): 403-413.

Riahi K, Rao S, Krey V, Cho C, Chirkov V, Fischer G, Kindermann G, Nakicenovic N, Rafaj P. 2011. RCP8.5: A scenario of comparatively high greenhouse gas emissions. Climatic Change, 109(1-2): 33-57.

Ritchie J T, Singh U, Godwin D C, Bowen W T. 1998. Cereal growth, development and yield. *In*: Tsuji G Y, Hoogenboom G, Thornton P K. Understanding Options for Agricultural Production. Dordrecht: Springer Netherlands: 79-98.

Ritchie J, Otter S. 1985. Description and performance of CERES-Wheat: a user-oriented wheat yield model. ARS-United States Department of Agriculture, Agricultural Research Service (USA). Wheat Yield Project, 38: 159-175.

Rodell M, Velicogna I, Famiglietti J S. 2009. Satellite-based estimates of groundwater depletion in India. Nature, 460(7258): 999-1002.

Roderick M L, Farquhar G D. 2004. Changes in Australian pan evaporation from 1970 to 2002. International Journal of Climatology, 24(9): 1077-1090.

Rosegrant M, Cai X, Cline S A. 2002. World water and food to 2025. International Food Policy Research Institute, Washington, DC. doi: 0-89629-646-6.

Rosenzweig C, Iglesias A, Yang X, Epstein P R, Chivian E. 2001. Climate change and extreme weather events; implications for food production, plant diseases, and pests. Global Change and Human Health, 2(2): 90-104.

Rosenzweig C, Jones J W, Hatfield J L, Ruane A C, Boote K J, Thorburn P, Antle J M, Nelson G C, Porter C, Janssen S, Asseng S. 2013. The agricultural model intercomparison and improvement project (AgMIP): protocols and pilot studies. Agricultural and Forest Meteorology, 170: 166-182.

Rosenzweig C, Parry M L. 1994. Potential impact of climate change on world food supply. Nature, 367(6459): 133-138.

Rosenzweig C, Tubiello F N, Goldberg R, Mills E, Bloomfield J. 2002. Increased crop damage in the US from excess precipitation under climate change. Global Environmental Change, 12(3): 197-202.

Rwasoka D, Gumindoga W, Gwenzi J. 2011. Estimation of actual evapotranspiration using the Surface Energy Balance System (SEBS) algorithm in the Upper Manyame catchment in Zimbabwe. Physics and Chemistry of the Earth, Parts A/B/C, 36(14): 736-746.

Sakthivadivel R, De Fraiture C, Molden D J, Perry C, Kloezen W. 1999. Indicators of land and water productivity in irrigated agriculture. International Journal of Water Resources Development, 15(1-2): 161-179.

Savage N. 2013. Modelling: predictive yield. Nature, 501(7468): S10-S11.

Schiermeier Q. 2011. Extreme measures. Nature, 477(7363): 148.

Semenov M A, Shewry P R. 2011. Modelling predicts that heat stress, not drought, will increase vulnerability of wheat in Europe. Scientific Reports, 1: 66.

Sheffield J, Wood E F, Roderick M L. 2012. Little change in global drought over the past 60 years. Nature, 491(7424): 435-438.

Shi W, Tao F, Zhang Z. 2013. A review on statistical models for identifying climate contributions to crop yields. Journal of Geographical Sciences, 23(3): 567-576.

Solomon S, Qin D, Manning M, Chen Z, Marquis M, Averyt K B, Tignor M, Miller H. 2007. Summary for Policymakers. *In*: Climate Change 2007: The Physical Science Basis. Contribution of Working Group I to the Fourth Assessment Report of the Intergovernmental Panel on Climate Change. Cambridge, UK: Cambridge University Press.

Stanhill G, Cohen S. 2001. Global dimming: a review of the evidence for a widespread and significant reduction in global radiation with discussion of its probable causes and possible agricultural consequences. Agricultural and Forest Meteorology, 107(4): 255-278.

Steltzer H, Post E. 2009. Seasons and life cycles. Science, 324(5929): 886-887.

Stöckle C O, Donatelli M, Nelson R. 2003. CropSyst, a cropping systems simulation model. European

Journal of Agronomy, 18(3): 289-307.

Su X, Singh V P, Niu J, Hao L. 2015. Spatiotemporal trends of aridity index in Shiyang River basin of northwest China. Stochastic Environmental Research and Risk Assessment, 29(6): 1571-1582.

Sun Q, Kröbel R, Müller T, Römheld V, Cui Z, Zhang F, Chen X. 2011. Optimization of yield and water-use of different cropping systems for sustainable groundwater use in North China Plain. Agricultural Water Management, 98(5): 808-814.

Teixeira A D C, Bastiaanssen W, Ahmad M D, Bos M. 2009. Reviewing SEBAL input parameters for assessing evapotranspiration and water productivity for the Low-Middle Sao Francisco River basin, Brazil: Part A: Calibration and validation. Agricultural and Forest Meteorology, 149(3): 462-476.

Thomas A. 2000. Spatial and temporal characteristics of potential evapotranspiration trends over China. International Journal of Climatology, 20(4): 381-396.

Thomas A. 2008. Agricultural irrigation demand under present and future climate scenarios in China. Global and Planetary Change, 60(3): 306-326.

Thorp K R, Hunsaker D J, French A N, White J W, Clarke T R, Pinter Jr P J. 2010. Evaluation of the CSM-CROPSIM-CERES-Wheat model as a tool for crop water management. Transactions of the ASAE (American Society of Agricultural Engineers), 53(1): 87.

Timsina J, Godwin D, Humphreys E, Kukal S, Smith D. 2008. Evaluation of options for increasing yield and water productivity of wheat in Punjab, India using the DSSAT-CSM-CERES-Wheat model. Agricultural Water Management, 95(9): 1099-1110.

Trenberth K E, Dai A, van Der Schrier G, Jones P D. Barichivich J, Briffa K R, Sheffield J. 2014. Global warming and changes in drought. Nature Climate Change, 4(1): 17-22.

Tuong T, Pablico P, Yamauchi M, Confesor R, Moody K. 2000. Increasing water productivity and weed suppression of wet seeded rice: effect of water management and rice genotypes. Experimental Agriculture, 36(01): 71-89.

Vicente-Serrano S M, Begueria S, Lopez-Moreno J I. 2011. Comment on "Characteristics and trends in various forms of the Palmer Drought Severity Index (PDSI) during 1900-2008" by Aiguo Dai. Journal of Geophysical Research-Atmospheres, 116 (D19): 112.

Wang F, Wang X, Ken S. 2004. Comparison of conventional, flood irrigated, flat planting with furrow irrigated, raised bed planting for winter wheat in China. Field Crops Research, 87(1): 35-42.

Wang H, Chen A, Wang Q, He B. 2015a. Drought dynamics and impacts on vegetation in China from 1982 to 2011. Ecological Engineering, 75: 303-307.

Wang H, Jia L, Steffen H, Wu P, Jiang L, Hsu H, Xiang L, Wang Z, Hu B. 2013a. Increased water storage in North America and Scandinavia from GRACE gravity data. Nature Geoscience, 6(1): 38-42.

Wang H, Vicente-serrano S M, Tao F, Zhang X, Wang P, Zhang C, Chen Y, Zhu D, El Kenawy A. 2016. Monitoring winter wheat drought threat in Northern China using multiple climate-based drought indices and soil moisture during 2000–2013. Agricultural and Forest Meteorology, 228: 1-12.

Wang J X, Huang J K, Yan T T. 2013b. Impacts of climate change on water and agricultural production in ten large river basins in China. Journal of Integrative Agriculture, 12(7): 1267-1278.

Wang J, Ma Y, Menenti M, Bastiaanssen W, Mitsuta Y. 1995. The scaling-up of processes in the heterogeneous landscape of HEIFE with the aid of satellite remote sensing. Journal of the Meteorological Society of Japan. Ser. II, 73(6): 1235-1244.

Wang W, Zhu Y, Xu R, Liu J. 2015b. Drought severity change in China during 1961–2012 indicated by SPI and SPEI. Natural Hazards, 75(3): 2437-2451.

Wang Y, Jiang T, Bothe O, Fraedrich K. 2007. Changes of pan evaporation and reference evapotranspiration in the Yangtze River basin. Theoretical and Applied Climatology, 90(1-2): 13-23.

Wheeler T, von Braun J. 2013. Climate change impacts on global food security. Science, 341(6145): 508-513.

Wilcox J, Makowski D. 2014. A meta-analysis of the predicted effects of climate change on wheat yields using simulation studies. Field Crops Research, 156: 180-190.

Wild M. 2014. Global dimming and brightening. Global Environmental Change: 39-47.

Wilhite D A, Svoboda M D, Hayes M J. 2007. Understanding the complex impacts of drought: A key to enhancing drought mitigation and preparedness. Water Resources Management, 21(5): 763-774.

Williams J R, Renard K G, Dyke P T. 1983. EPIC: A new method for assessing erosion's effect on soil productivity. Journal of Soil and Water Conservation, 38(5): 381-383

Wu S, Luo Y, Wang H, Gao J, Li C. 2016. Climate change impacts and adaptation in China: Current situation and future prospect. Chinese Science Bulletin, 61(10): 1042-1054.

Xiong W, Conway D, Lin E, Holman I. 2009. Potential impacts of climate change and climate variability on China's rice yield and production. Climate Research, 40(1): 23-35.

Xu K, Yang D, Yang H, Li Z, Qin Y, Shen Y. 2015. Spatio-temporal variation of drought in China during 1961–2012: A climatic perspective. Journal of Hydrology, 526: 253-264.

Yan N, Wu B. 2014. Integrated spatial-temporal analysis of crop water productivity of winter wheat in Hai Basin. Agricultural Water Management, 133: 24-33.

Yang J M, Yang J Y, Dou S, Yang X M, Hoogenboom G. 2013b. Simulating the effect of long-term fertilization on maize yield and soil C/N dynamics in northeastern China using DSSAT and CENTURY-based soil model. Nutrient Cycling in Agroecosystems, 95(3): 287-303.

Yang J Y, Qin L, Mei X R, Yan C R, Hui J U, Xu J W. 2013a. Spatiotemporal characteristics of reference evapotranspiration and its sensitivity coefficients to climate factors in Huang-Huai-Hai Plain, China. Journal of Integrative Agriculture, 12(12): 2280-2291.

Yang J, Mei X, Huo Z, et al., 2015. Water consumption in summer maize and winter wheat cropping system based on SEBAL model in Huang-Huai-Hai Plain, China. Journal of Integrative Agriculture, 14(10): 2065-2076.

Yang Y, Yang Y, Moiwo J P, Hu Y. 2010. Estimation of irrigation requirement for sustainable water resources reallocation in North China. Agricultural Water Management, 97(11): 1711-1721.

Yong B, Ren L, Hong Y, Gourley J J, Chen X, Dong J, Wang W, Shen Y, Hardy J. 2013. Spatial–temporal changes of water resources in a typical semiarid basin of North China over the past 50 years and assessment of possible natural and socioeconomic causes. Journal of Hydrometeorology, 14(4): 1009-1034.

Yu M, Li Q, Hayes M J, Svoboda M D, Heim R R. 2014a. Are droughts becoming more frequent or severe in China based on the Standardized Precipitation Evapotranspiration Index: 1951-2010?

International Journal of Climatology, 34(3): 545-558.

Yu Q, Li L, Luo Q, Eamus D, Xu S, Chen C, Wang E, Liu J, Nielsen D C. 2014b. Year patterns of climate impact on wheat yields. International Journal of Climatology, 34(2): 518-528.

Yuan B, Gou J P, Ye M Z, Zhao J F. 2012. Variety distribution pattern and climatic potential productivity of spring maize in Northeast China under climate change. Atmospheric Science, 57(14): 1252-1262.

Zhai P, Zou X. 2005. Changes of temperature and precipitation and their effects on drought in China during 1951–2003. Advances in Climate Chang Research, 1(1): 16-18.

Zhang H L, Zhao X, Yin X G, Liu S L, Xue J F, Wang M, Pu C, Lal R, Chen F. 2015. Challenges and adaptations of farming to climate change in the North China Plain. Climatic Change, 129(1-2): 213-224.

Zhang H, Wang X, You M, Liu C. 1999. Water-yield relations and water-use efficiency of winter wheat in the North China Plain. Irrigation Science, 19(1): 37-45.

Zhang K, Kimball J S, Running S W. 2016. A review of remote sensing based actual evapotranspiration estimation. Wiley Interdisciplinary Reviews: Water, 3(6): 834-853.

Zhang Y, Yu Q, Liu C, Jiang J, Zhang X. 2004. Estimation of winter wheat evapotranspiration under water stress with two semiempirical approaches. Agronomy Journal, 96(1): 159-168.

Zhao R F, Chen X P, Zhang F S, Zhang H, Schroder J, Römheld V. 2006. Fertilization and nitrogen balance in a wheat–maize rotation system in North China. Agronomy Journal, 98(4): 938-945.

Zheng Z, Cai H, Yu L, Hoogenboom G. 2016. Application of the CSM-CERES-Wheat Model for yield prediction and planting date evaluation at Guanzhong Plain in Northwest China. Agronomy Journal, 109: 204-217.

Chapter 2 Impacts of climate change on potential evapotranspiration under a historical period and future climate scenario in the Huang-Huai-Hai Plain, China

Abstract

Climate change is widely accepted to be one of the most critical problems faced by the Huang-Huai-Hai Plain, which is a region in where groundwater is over-exploited, and future warmer and drought conditions might intensify crop water demand. In this study, the spatio-temporal patterns of ET_0 and primary driving meteorological variables were investigated based on a historical period and RCP8.5 scenario daily data set from 40 weather stations over the Huang-Huai-Hai Plain using linear regression, spline interpolation method, a partial derivative analysis, and multivariate regression. The results indicated a negative trend in all the analysis periods (except spring) of the past 54 years, and the trends of only summer and the entire year were statistically significant ($P<0.01$) with slopes of -1.09 mm·y^{-1} and -1.29 mm·y^{-1} respectively. In contrast, a positive trend was observed in all four seasons and the entire year under the RCP8.5 scenario, with the biggest increment equal to 1.36 mm·y^{-1} in summer and an annual increment of 3.37 mm·y^{-1}. The spatial patterns of the seasonal and annual ET_0 exhibited the lowest values in southeastern regions and the highest values in northeastern parts of Shandong Province, probably because of the combined effects of various meteorological variables over the past 54 years. Relative humidity (RH) together with solar radiation (RS) were detected to be the main climatic factors controlling the reduction of ET_0 in summer, autumn, and the entire year in the Huang-Huai-Hai Plain. ET_0 in spring was mainly sensitive to changes in RS and RH, whereas ET_0 in winter was most sensitive to changes in wind speed (WS) and decreased due to declining RH. Under the future RCP8.5

scenario, the annual ET_0 distribution displays a rich spatial structure with a clear northeast-west gradient and an area with low values in the southern regions, which is similarly detected in spring and summer. The most sensitive and primary controlling variables with respect to the increment of future ET_0 are RS and then mean temperature in spring, while turn to be mean temperature and then RS in summer. In autumn, future ET_0 is most sensitive to RH changes. WS and RH are the controlling variables for ET_0 in winter. Future annual ET_0 is most sensitive to RH changes and accordingly RS is responsible for the predicted increment of the annual ET_0. The better understanding of current and future spatio-temporal patterns of ET_0 and the regional response of ET_0 to climate change can contribute to the establishment of a policy to realize more efficient use of water resources and sustainable agricultural production in the Huang-Huai-Hai Plain.

2.1 Introduction

Potential evapotranspiration (ET_0) is widely acknowledged as a key hydrological variable representing an important water loss from catchments. It is closely related to groundwater recharge, runoff generation, and soil water movements, some important terms of hydrological processes (Xu and Singh, 2005; Zhang et al., 2011), and with which it can be used to estimate actual evapotranspiration (ET_a), schedule irrigation and other agricultural management practices (Dyck, 1985; Hobbins et al., 2001; Xu and Li, 2003). Drought conditions are anticipated to be aggravated due to climate change by increasing potential evapotranspiration and augmenting crop water consumption in water-limited regions (Goyal, 2004; Maracchi et al., 2005; Thomas, 2008). It has been recognized that an in-depth understanding of the spatio-temporal pattern of evapotranspiration and an accurate estimation of its response to climate change are essential to efficient water resources management and water productivity assessment. (Chen et al., 2007a; Drogue et al., 2004; Han et al., 2014; Yang et al., 2015).

According to the predictions of climate change models, ET_0 is expected to increase over the coming years due to expected temperature rise (Goyal, 2004; McNulty et al., 1997). However, decreasing trends of ET_0 have been detected in some regions of China (Chen et al., 2006; Thomas, 2000; Wang et al., 2007), Korea (Song et al., 2014), the United States (Suat et al., 2012), and Australia (Roderick and Farquhar, 2004). Therefore, global atmospheric temperatures rise might not necessarily give rise to ET_0 in all cases. For example, the reduction in solar radiation could compensate for the impact of temperatures on ET_0 as depicted in many places (Stanhill and Cohen, 2001;

Wild, 2014). Chen et al. (2006) concluded that the reduction of WS was the primary meteorological variable contributing to the observed decline of ET_0 rates on the Tibetan Plateau. While, Gao et al. (2006) and Wang et al. (2007) have identified reductions in WS and RS as the primary contributing factors to decreasing trends of ET_0. Atmospheric temperature is probably thought to be the most widely used marker of climatic change on both regional and global scales. According to the 2013 IPCC report, over the past 100 years (1913–2012), global temperature has risen by 0.91°C (Stocker et al., 2014) and is expected to continue to rise throughout the 21st century, altering the hydrological cycle by affecting precipitation or/and evaporation (Huntington, 2006). The patterns of the spatio-temporal distribution of climatic variables in response to global warming remain a matter of active debate, and related studies are increasing worldwide (Feddema, 1999; Georgia, 2000; McNulty et al., 1997; Muhs and Maat, 1993). Consequently, analysis of the contribution of the main four climate factors to in the estimated ET_0 is essential.

A sensitivity analysis is essential to capturing the importance of the meteorological variables in the Penman-Monteith formulation, and ascertain the effects of climate change on ET_0 (Zhang et al., 2010). Several researches have described the sensitivity of ET_0 to meteorological factors in selected climate scenarios (Goyal, 2004; Irmak et al., 2006; Rana and Katerji, 1998). Furthermore, the most effective variables have been reported to be wind speed (Cohen et al., 2002; Todisco and Vergni, 2008; Wang et al., 2007), solar radiation (Gao et al., 2006; Wang et al., 2007) and relative humidity (Gong et al., 2006). Song et al. (2009) indicated that for entire North China plain annual ET_0 has undergone a statistically pronounced reduction of 11.9 mm·decade^{-1} in the past 46 years and that reductions of RS and WS have exerted a greater influence on ET_0 than the other variables. Although analyses of the sensitivity of ET_0 to climatic factors have been performed before, a focus on the ranges of variation of the variables in the historic record of the Huang-Huai-Hai Plain is not necessarily all covered. In addition, few studies have focused on future changes of ET_0 (especially in the Huang-Huai-Hai Plain), which limits the potential of the former studies to provide guidelines for more efficient water resources management. To predict future changes of ET_0, climate scenarios are often used to provide future climatic factors for the estimation. A few studies have calculated ET_0 directly using climatic factors from such climate scenarios. In order to obtain information on future climate change with higher spatial and temporal resolutions, most studies have generated future ET_0 by the downscaling General Circulation Models outputs. Li et al. (2012)investigated the spatio-temporal variability of ET_0 during 1961–2009 and 2011–2099 under two emission scenarios (A2 and B2) on the Loess Plateau of China. Zhou and Hong (2013) investigated the projected changes of the Palmer drought severity index (PDSI) with ET_0 under the

Representative Concentration Pathway 8.5 (RCP8.5). This RCP8.5 scenario can be considered as a worst-case scenario since it represents future high CO_2-emission rates (increase by about 120 Gt CO_2-eq. by 2100 compared to 2000).

The Huang-Huai-Hai Plain is a major crop-producing area in China that encompasses 18 million hectares of intensively farmed arable land (19% of the country's total crop producing land). Water is thought to be the main factor limiting agricultural production in the Huang-Huai-Hai Plain. Furthermore, water limitations will probably be accentuated by continuous food demand, soil quality deterioration and over-exploitation of groundwater resources (Chen et al., 2003). Moreover, changes in climate have intensified with annual precipitation decreasing at an average rate of 2.92 mm·y^{-1} (Liu et al., 2010). Thus, the regional aggravation of water scarcity has led to an increased awareness of the importance of water as an agricultural resource for crop production in the Huang-Huai-Hai Plain.

Given this context, the research presented in this chapter had the following objectives: (1) An investigation of the characteristics of the meteorological variables and ET_0 on seasonal and annual scales over the Huang-Huai-Hai Plain under a historical (1961–2014) period and the RCP8.5 scenario, using linear regression and spline interpolation; (2) The establishment of a sensitivity analysis for ET_0 on seasonal and annual scales using the partial derivative method; (3) To identify the primary variables controlling changes in ET_0 in the context of climate change.

2.2 Materials and methods

2.2.1 Study area

The Huang-Huai-Hai Plain (31°14′–40°25′N, 112°33′–120°17′E) is known as the largest agricultural production region in China. The climate is temperate, sub-humid, and continental monsoon, with average annual precipitation of 500–800 mm (Ren et al., 2008). Annual precipitation is concentrated in summer (July to September) and winter is strongly characterized by a lack of water for agricultural production. Water shortage in the Huang-Huai-Hai Plain have become of considerable concern over recent decades (Brown and Rosenberg, 1997). The plain encompasses around 18 million hectares of farmland of which about 61% and 31% are dedicated to wheat and maize production, respectively (He et al., 2009). The main cropping system is well known as the rotation of wheat and maize with the systematic application of irrigation water and fertilizer in the Huang-Huai-Hai Plain (Liang et al., 2011; Sun et al., 2011; Zhao et al., 2006).

Table 2-1　The information of meteorological stations over the Huang-Huai-Hai Plain, China

Station number	Location	Province	Longitude	Latitude
58015	Tangshan	Anhui	116°19'	34°25'
58102	Bozhou	Anhui	115°46'	33°52'
58122	Su Xian	Anhui	116°59'	33°37'
58203	Fuyang	Anhui	115°49'	33°38'
58215	Shou Xian	Anhui	116°47'	32°33'
58221	Bengbu	Anhui	117°23'	32°57'
58311	Liuan	Anhui	116°30'	31°45'
58321	Hefei	Anhui	117°14'	31°52'
54511	Beijing	Beijing	116°17'	39°56'
53698	Shijiazhuang	Hebei	114°25'	38°2'
53798	Xingtai	Hebei	114°30'	37°4'
54518	Langfang	Hebei	116°23'	39°40'
54534	Tangshan	Hebei	118°9'	39°40'
54539	Laoting	Hebei	118°31'	39°25'
54602	Baoding	Hebei	115°31'	38°51'
54606	Raoyang	Hebei	115°43'	38°14'
54624	Huanghua	Hebei	117°20'	38°22'
54705	Nangong	Hebei	115°23'	37°22'
53898	Anyang	Henan	114°22'	36°7'
53986	Xinxiang	Henan	113°53'	35°19'
57083	Zhengzhou	Henan	113°51'	34°43'
57089	Xuchang	Henan	113°51'	34°1'
57091	Kaifeng	Henan	114°23'	34°46'
57193	Xihua	Henan	114°31'	33°47'
57290	Zhumadian	Henan	114°1'	33°
58005	Shangqiu	Henan	115°40'	34°27'
58208	Gushi	Henan	115°40'	32°10'
58027	Xuzhou	Jiangsu	117°9'	34°17'
58040	Ganyu	Jiangsu	119°7'	34°50'
58138	Yutai	Jiangsu	118°31'	32°59'
54725	Huimin	Shandong	117°32'	37°30'
54808	Chaoyang	Shandong	115°35'	36°2'
54826	Taishan	Shandong	117°6'	36°15'
54843	Weifang	Shandong	119°5'	36°42'
54916	Yanzhou	Shandong	116°51'	35°34'
54945	Rizhao	Shandong	119°32'	35°22'
54527	Tianjin	Tianjin	117°10'	39°6'
54623	Tanggu	Tianjin	117°43'	39°

2.2.2 Meteorological data

A historical dataset from 1961 to 2014, composed of data from 40 meteorological stations, was provided by the China Meteorological Administration (Table 2-1). Daily maximum (T_{max}, °C) and minimum temperatures (T_{min}, °C), average relative humidity (RH, %), wind speed (WS, m·s^{-1}) observed at 10 m height, and daily sunshine duration (SD, h) data were included. The 0.5° × 0.5° gridded data of the Huang-Huai-Hai Plain from 2015 to 2099 simulated under the future RCP8.5 climatic scenario and obtained from the National Climate Center, included daily average temperature (T_a, °C), daily highest (T_{max}, °C) and lowest temperature (T_{min}, °C), daily precipitation (P, mm), daily average wind speed (WS, m·s^{-1}), daily average relative humidity (RH, %), and daily

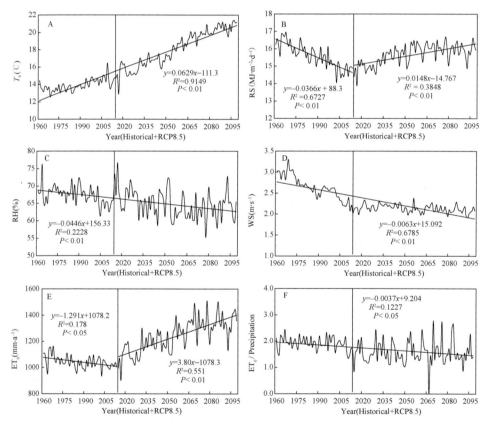

Figure 2-1 Annual (A) daily surface air temperature (T_a), (B) daily net radiation (RS), (C) daily average relative humidity (RH), (D) daily average wind speed (WS), (E) potential evapotranspiration (ET$_0$) and (F) ET$_0$/precipitation in the historical period and under RCP8.5 scenario. The data were obtained from area-averaged values of meteorological variables, ET$_0$ and ET$_0$/precipitation from 40 stations in the past 54 years and future 85 years

net radiation (RS, MJ·m^{-2}·d^{-1}). The RCP8.5 scenario is characterized by high concentration of greenhouse gas with stabilizing emissions post-2099 (increase by about 120 Gt CO_2-eq. by 2100 compared to 2000) (Riahi et al., 2011). We used a bilinear interpolation method to interpolate the RCP8.5 gridded data to the stations, and the results were then validated by the observational data during 2005–2014. The annual temperature, net radiation, average relative humidity, and wind speed in the historical period and under RCP8.5 scenario are plotted in Figure 2-1. A remarkable positive trend with a slope of 0.63°C·decade^{-1} can be seen for T_a, whereas WS and RH have significant negative trends with slopes of –0.06 m·s^{-1}·decade^{-1} and –0.45%·decade^{-1}, respectively. For RS, a slope of –0.37 MJ·m^{-2}·d^{-1}·decade^{-1} is observed based on the recorded dataset in the historical period, whereas a significant positive trend with a slope of 0.15 MJ·m^{-2}·d^{-1}·decade^{-1} can be seen based on predicted dataset under the RCP8.5 scenario.

2.2.3 Estimation of potential evapotranspiration

The Penman-Monteith formula recommended by the Food and Agriculture Organization (FAO) of the United Nations was used to calculate the ET_0 for both historical and future conditions:

$$ET_0 = \frac{0.408\Delta(R_n - G) + \gamma \frac{900}{T+273} U_2(e_s - e_a)}{\Delta + \gamma(1 + 0.34 U_2)} \quad \text{Formula 2-1}$$

where, ET_0 is potential evapotranspiration (mm·d^{-1}); Δ represents the slope of the saturation vapor pressure/temperature curve (kPa·°C^{-1}); R_n is the net radiation from the canopy (MJ·m^{-2}·d^{-1}); G is the soil heat flux (MJ·m^{-2}·d^{-1}); T_a is the daily temperature; U_2 is the wind velocity (m·s^{-1}); e_s is the saturation vapor pressure (kPa); e_a is the actual water vapor pressure (kPa); $e_s - e_a$ is the vapor pressure deficit (kPa); γ is the psychometric constant (kPa·°C^{-1}). Allen et al. (1998) explain how to obtain the parameters and variables from meteorological data.

2.2.4 Time series analysis to quantify major trends

Both parametric and non-parametric methods have been widely used to identify trends in climatic variables. However, recent studies suggest that non-parametric tests are more suitable for non-normally distributed and censored data, which are frequently encountered in meteorological and hydrological time series. Among them, the non-parametric Mann-Kendall test (Kendall, 1975; Mann, 1945) is widely accepted to be a popular method for trend analysis of hydrological and meteorological series (Dinpashoh et al., 2011; Jhajharia et al., 2012; Kousari and Ahani, 2012; Zheng and

Wang, 2014) with little influence by the presence of outliers in the data. In addition, it is highly recommended for general use by the World Meteorological Organization (Zhang et al., 2011). Therefore, we used the Mann-Kendall test for the trend analysis of our data. Furthermore, the magnitude of the trends in the time series was estimated using the non-parametric Theil-Sen's slope (Sen, 1968; Theil, 1992), which is robust because it avoids the effect of outlier values (Jhajharia et al., 2015; Su et al., 2015; Xu et al., 2003; Yue et al., 2002). The confidence level used in this study is 95%. In our study, we applied two Mann-Kendall tests recommended in the literature (Kumar et al., 2009; Nalley et al., 2012; Zamani et al., 2016): (1) a MK test considering all significant autocorrelation structures (MK3) and (2) a MK test considering long-term persistence (LTP) and Hurst coefficient (MK4). The two modified Mann-Kendall tests and Theil-Sen's slope estimator involved in this study were performed using R software version 3.2.4 (Team, 2011).

2.2.5 Sensitivity analysis and multivariate regression

McKenney and Rosenberg (1993) and Goyal (2004) have carried out a sensitivity analysis for ET_0 where the variations of output ET_0 were plotted against the relative changes of input meteorological factors. Irmak et al. (2006) then introduced the sensitivity coefficients on a daily step by dividing the amount of increase/decrease in ET_0 by the increase/decrease of each meteorological factor. However, Zhang et al. (2010) highlighted that such sensitivity coefficient is susceptible to the effects of variations in the magnitudes of ET_0 and the meteorological factors. A non-dimensional sensitivity coefficient can be normalized as

$$SC = \frac{\partial ET_0}{\partial V} = \frac{\Delta ET_0 / ET_0}{\Delta CV / CV} \qquad \text{Formula 2-2}$$

where, SC is the dimensionless sensitivity coefficient, and ET_0 and CV (air temperature, relative humidity, solar radiation or wind speed) are base values before the change.

To investigate the major controlling variables of ET_0 trends, multivariate regression analysis was conducted with the meteorological factors in the Huang-Huai-Hai Plain as the controlling climate factors (Chun et al., 2012; Ma et al., 2012).

2.3 Results

2.3.1 Historical and future trends of meteorological variables

Seasonal trends of T_a, RS, RH and WS during 1961–2014 and under the RCP8.5 scenario of 2015–2099 are presented in Table 2-2. The result of Mann-Kendall test of

T_a is characterized by an increasing trend both in the historical period and under the RCP8.5 scenario. The trend of increase in T_a is 0.41°C·decade^{-1} and 0.39°C·decade^{-1} for spring and winter, respectively, over the past 54 years. Similarly, in autumn, a significantly increasing trend of 0.23°C·decade^{-1} is detected. The increase in T_a is higher in the cooler seasons than in the warmer seasons. Furthermore, T_a increases by 0.64°C·decade^{-1} and 0.80°C·decade^{-1} for spring and winter, respectively, under the future RCP8.5 condition. Similarly, in summer and autumn, a significantly increasing trend of 0.76°C·decade^{-1} and 0.77°C·decade^{-1} is also detected. Accordingly, the increase in T_a is higher under the RCP8.5 scenario than for the past 54 years.

Table 2-2 Trend analysis of meteorological variables with Mann-Kendall test and Theil-Sen's slope estimator under the historical period and RCP8.5 scenarios

	T_a (°C)	RS (MJ·m^{-2})	RH (%)	WS (m·s^{-1})
Historical				
Spring	0.41*	−23.3**	−1.28*	−0.19**
Summer	0.07*	−65.7**	−0.49*	−0.12**
Autumn	0.23**	−28.9**	−1.02**	−0.16**
Winter	0.39**	−23.9*	−0.60	−0.20**
Entire year	0.28**	−139.3**	−0.73**	−0.17**
RCP8.5				
Spring	0.64**	12.6**	−0.09	−0.021**
Summer	0.76**	21.7**	−0.81**	−0.014*
Autumn	0.77**	16.9**	−0.47	−0.022**
Winter	0.80**	1.58	−0.12	−0.012**
Entire year	0.77**	53.1**	−0.38*	−0.018**

Notes: The values described in the table were obtained from area-averaged values of meteorological variables from 40 meteorological stations using Theil-Sen's slope estimator in the past 54 years and future 85 years. In Table 2-2 and Table 2-3, the positive values mean upward trends and negative values mean downward trends. No star means no significant trend ($P>0.05$); * indicates trend ($P<0.05$), and ** indicates a significance level of 0.01. The same below

For RS in the past 54 years, a significant negative trend is detected in all seasons and during the entire year, i.e., 139.3 MJ·m^{-2}·decade^{-1}, whereas under the RCP8.5 scenario, an increasing trend is observed in all analysis seasons, only one (winter) of which is not statistically significant. It is noted that RH has a significant decreasing trend over the past 54 years. In spring and winter, RH reduces by 1.28%·decade^{-1} and 0.73%·decade^{-1}, respectively. Similarly, it reduces by 0.49%·decade^{-1} and 1.02%·decade^{-1} in summer and autumn, respectively. Under the RCP8.5 scenario, a negative trend can be observed in all analysis seasons, only one (summer) of which is statistically significant. In all seasons, WS shows a negative trend and the slope can be observed to be slightly great over the past 54 years than under the RCP8.5 scenario. In general, for annual RH and

WS, significant declining trends of $-0.77\%\cdot\text{decade}^{-1}$ and -0.17 m·s^{-1}·decade^{-1} can be found over the past 54 years, which are slightly stronger than under the RCP8.5 scenario. The seasonal-scale changes of T_a over the past 54 years and under the RCP8.5 scenario show significant increasing trends of $0.28\%\cdot\text{decade}^{-1}$ and $0.75\%\cdot\text{decade}^{-1}$, respectively.

2.3.2 Spatial and temporal characteristics of ET$_0$

The investigation of the trends of ET$_0$ and their persistence using modified Mann-Kendall test and Theil-Sen's slope estimator during the historical period and under the RCP8.5 scenario gives a better insight in the expected seasonal differences of water demand and the evolution of these demands in the future. As shown in Table 2-3, average ET$_0$ is highest in summer, accounting for 39.4% and 38.5% of the annual ET$_0$ under the historical period and RCP8.5 scenario, respectively, followed in descending order by spring, autumn, and winter. Over the past 54 years, a negative trend occurs in all analysis periods (except spring), whereas of which statistically significant slopes of -1.08 mm·y^{-1} and -1.20 mm·y^{-1} are detected in summer and the entire year. In contrast, a positive trend can be observed in all four seasons and the entire year under the RCP8.5 scenario; the biggest seasonal increment is 1.35 mm·y^{-1} in summer and the annual increment is 3.45 mm·y^{-1}. On average, the prediction of ET$_0$ (RCP8.5 scenario) is larger than what was observed during the past 54 years: 1042.4 mm for the historical data set and 1230.3 mm for the RCP8.5 scenarios.

Table 2-3 Trend analysis of ET$_0$ with Mann-Kendall test and Theil-Sen's slope estimator under the historical period and RCP8.5 scenarios

	Spring		Summer		Autumn		Winter		Entire year	
	Value (mm)	Slope (mm·y^{-1})	Value (mm)	Slope (mm·y^{-1})	Value (mm)	Slope (mm·y^{-1})	Value (mm)	Slope (mm·y^{-1})	Value (mm)	Slope (mm·y^{-1})
Historical	322.5	0.08	410.4	-1.08^*	206.4	-0.21	102.8	-0.11	1042.4	-1.20^{**}
RCP8.5	388.5	0.82^{**}	473.4	1.35^{**}	255.1	0.85^{**}	113.3	0.35^{**}	1230.3	3.45^{**}

Notes: The values described in the table were obtained from area-averaged ET0 from 40 meteorological stations using Theil-Sen's slope estimator in the past 54 years and future 85 years

Seasonal and annual spatial variability of ET$_0$ can be observed during the historical period and under the RCP8.5 scenario. In the historical dataset of 1961–2014 for the Huang-Huai-Hai Plain, the regional difference ranges from 652.1 mm·y^{-1} to 1182.0 mm·y^{-1}. The coefficient of variation for the average annual ET$_0$ of the 40 stations is 0.10, which signifies rather low spatial variation. Figure 2-2 shows the spatial patterns for the ET$_0$ in four seasons and the entire year in the Huang-Huai-Hai Plain during 1961–2014. The lowest ET$_0$ values (< 950 mm) are found in a small region of the southeast.

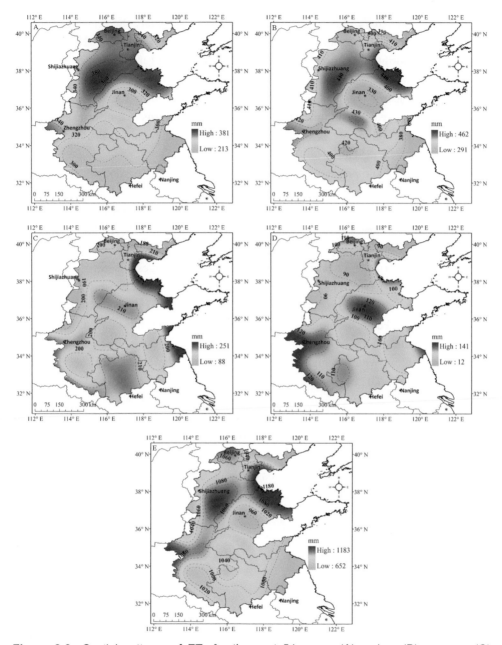

Figure 2-2 Spatial patterns of ET_0 for the past 54 years: (A) spring, (B) summer, (C) autumn, (D) winter and (E) the entire year. These maps were obtained using the Spline method for interpolation with Geostatistical Analysis module in the software ArcGIS

The annual T_a and RH decrease from the southeast to the northwest, whereas the annual WS and RS increase from the southeast to the northwest (Liu et al., 2013a; Yang et al.,

2011). Low values of T_a, WS, and RS and high values of RH are the primary causes of lowest ET_0 values in the southeastern region. The highest ET_0 values (> 1000 mm) are found in the northeast region of Shandong Province and south region of Hebei Province, which can mainly be attributed to the low RH, high WS and strong RS. The ET_0 characteristics of the four seasons (except summer) follow the spatial pattern of ET_0 for the entire year. In summer, the spatial distribution turns to a different pattern in which the highest values are observed in the northeast of Shandong Province and the south part of the Huang-Huai-Hai Plain.

Figure 2-3 shows the projected (RCP8.5) spatial patterns of seasonal and annual ET_0 for the Huang-Huai-Hai Plain. The distribution of the annual averages has a heterogeneous spatial structure including a northeast-west gradient of relatively low ET_0 areas and high areas in the southern region, similar patterns are found in spring and summer. A spatial comparison of ET_0 average predicted under the RCP8.5 with respect to the historical ET_0 averages reveals where the changes are expected to be the largest in the Huang-Huai-Hai Plain. We observed the largest increase with respect to the past in the southwest by a change in magnitude of 25%–32%, whereas the smallest increase was located in the northeast (3%–10%). As for autumn and winter, marked areas of low values are visible in the projected pattern of ET_0 from the north to central parts of the region, whereas higher values can be found in eastern and southwestern regions.

The results from the investigation of ET_0 trends at annual time scale for the past data and the RCP8.5 scenario are shown in Figure 2-4 in order to compare the results of each station according to its location in the Huang-Huai-Hai Plain. We also compared the two versions of the MK test (MK3/MK4). Nearly 12.5% and 32.5% of the stations witnessed significantly decreasing trends at the 1% level of significance using MK3 and MK4 tests, respectively, while these percentages changed to 50% and 55%

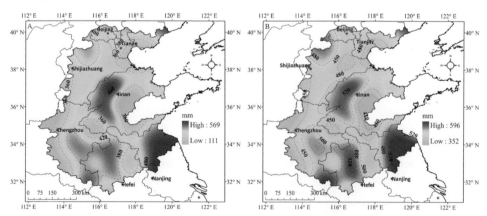

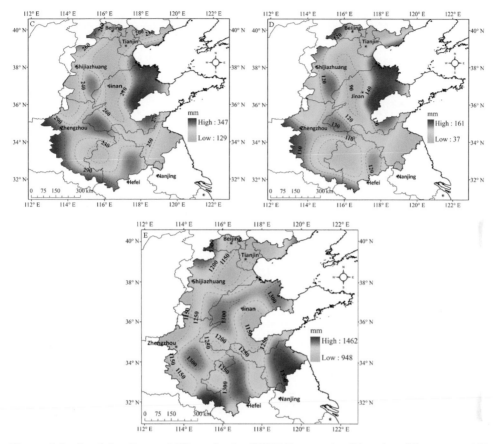

Figure 2-3 Spatial patterns of ET_0 under the RCP8.5 scenario: (A) spring, (B) summer, (C) autumn, (D) winter and (E) the entire year. These maps were obtained using the Spline method for interpolation with Geostatistical Analysis module in the software ArcGIS

for the past 54 years. Additionally, for the RCP8.5 scenario, the percentage of significantly increasing trends was 100% for both MK3 and MK4 tests. Comparing the results of the two versions of the Mann-Kendall test, we found that the Z-statistic values of MK3 test were slightly higher than those obtained from the MK4 test, indicating that the consideration of Hurst coefficient has led to a decrease in the level of significance, in accordance with the findings in streamflow and precipitation for MK3 and MK4 tests in Iran (Zamani et al., 2016). As already mentioned, the historical data set mainly displays annually decreasing trends, whereas the RCP8.5 scenario results in increasing trends in general. As for the spatial variability of ET_0, a significant increase is consistent all over the plain and stations with relative greater increment mainly in the southwest, and some in the north.

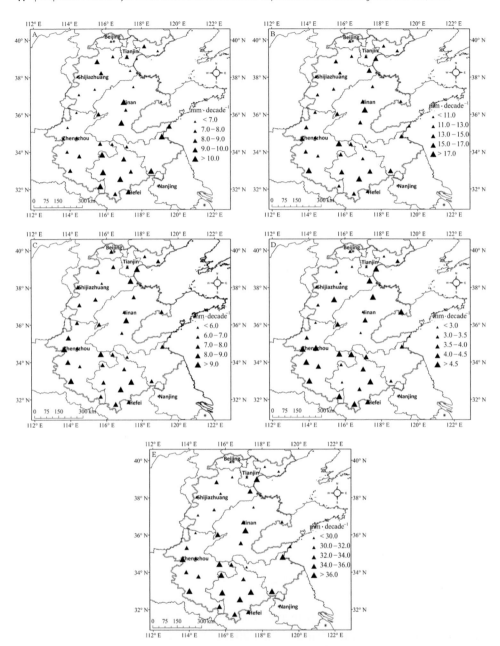

Figure 2-4 Spatial patterns of ET_0 trends under the RCP8.5 scenario: (A) spring, (B) summer, (C) autumn, (D) winter, and (E) the entire year. These maps were obtained with the software ArcGIS with upward arrows for the ascending trend

2.3.3 Temporal variation of sensitivity coefficients

Table 2-4 lists annual and seasonal trends of sensitivity coefficients for ET_0 during historical period and under the RCP8.5 scenario. It shows that the most effective meteorological factor impacting ET_0 varied with season and scenario. For the historical period, ET_0 is most influenced by solar radiation (positive relationship) and relative humidity (negative relationship). Only during winter, wind speed becomes the most influential factor. On a yearly basis, sensitivity coefficient for solar radiation (S_{RS}) equals 0.4 and sensitivity coefficient for relative humidity (S_{RH}) equals –0.93. The overall trend of the influence of T and RS on ET_0 is negative (negative slope of their sensitivity coefficients over time), whereas RH and WS are of increasing importance (positive slope of their sensitivity coefficients over time). Looking at the results of the RCP8.5 scenario, we get a different view. Solar radiation is still the variable having a positive impact on potential evapotranspiration and this relationship becomes more important as we go through the future scenario. Looking at yearly averages, relative humidity has the highest absolute value of sensitivity coefficient and a negative relationship with respect to ET_0; whereas solar radiation is the most important factor positively influencing ET_0. The trends of the sensitivity coefficients of RS, RH and WS are all positive, only the influence of temperature seems to decrease with time in the future scenario.

Table 2-4 Trend analysis of sensitivity coefficients for ET_0 with Mann-Kendall test and Theil-Sen's slope estimator during the historical period and under the RCP8.5 scenario

	S_T		S_{RS}		S_{RH}		S_{WS}	
	value	slope (y^{-1})	value	slope (y^{-1})	value	slope (y^{-1})	value	slope (y^{-1})
Historical								
Spring	–0.22	–0.016**	0.45	0.000	–0.44	0.050**	0.14	0.010**
Summer	–0.62	–0.012**	0.66	–0.005**	–0.69	0.001	0.07	0.007**
Autumn	–0.27	–0.012**	0.37	–0.005**	–0.53	0.033*	0.18	0.009**
Winter	0.08	–0.013	0.12	–0.003*	–0.03	0.002**	0.27	0.007*
Entire year	–0.26	–0.014**	0.40	–0.003**	–0.93	0.006*	0.17	0.008**
RCP8.5								
Spring	–0.36	–0.026**	0.45	0.004**	–0.44	0.042**	0.18	0.003*
Summer	–0.90	–0.048**	0.58	0.004*	–0.07	0.007**	0.18	0.005**
Autumn	–0.43	–0.033**	0.36	0.007**	–0.62	0.071**	0.23	0.003
Winter	–0.03	–0.016**	0.16	0.001	–0.04	0.006**	0.22	0.008**
Entire year	–0.43	–0.031**	0.39	0.004**	–0.65	0.002**	0.20	0.005**

Notes: The values described in the table were obtained from area-averaged values of sensitivity coefficients for ET_0 from 40 meteorological stations using Theil-Sen's slope estimator in the past 54 years and future 85 years

As described in Table 2-4, the most effective meteorological factor impacting ET_0 varied with season and scenario. For the historical period, the ET_0 in spring is most sensitive to RS (S_{RS}=0.45), and in winter to WS (S_{WS}=0.27). Trends of S_{RS} are negative in the time series analysis, indicating that the negative influence on ET_0 became weaker in the past 54 years, in combination with the positive value of S_{RS}. However, the positive influence on ET_0 strengthened in combination with the positive value of S_{WS}. The ET_0 in summer, autumn, and the entire year shows greatest sensitivity to RH, i.e., the sensitivity coefficients of RH (S_{RH}) in summer, autumn, and entire year (−0.69, −0.53, and −0.93, respectively) are larger than for the other variables. Trends of S_{RH} are positive in the time series analysis, indicating that the negative influence on ET_0 became smaller in the past 54 years, in combination with the negative value of S_{RH}.

Similar to the historical condition, the ET_0 in spring and winter under the RCP8.5 scenario shows greatest sensitivity to RS (S_{RS}=0.45) and WS (S_{WS}=0.22). Trends of S_{RS} and S_{WS} are positive in the time series analysis, indicating that the positive influence on ET_0 strengthens in the future 74 years, in combination with their positive values. The ET_0 in autumn and the entire year shows greater sensitivity to RH (S_{RH}=−0.62 and −0.65, respectively) than to the other variables. Trends of S_{RH} in the time series analysis indicate that the negative influence on ET_0 become weaker. An average temperature is the most sensitive factor to ET_0 in summer (S_T=−0.90) and its negative influence on ET_0 strengthens in the future 74 years. Figure 2-5 shows the spatial patterns of these ET_0 trends under the RCP8.5 scenario.

In this study, sensitivity centers of ET_0 were calculated using a gravity center analysis method, based on the data-set of sensitivity coefficients for the 40 stations under the historical and future conditions, in order to investigate better the characteristics and spatial differentiation of the sensitivity coefficients for ET_0 (Gaile, 1984). Figure 2-6 shows that the centers of S_T, S_{RS}, S_{WS} and S_{RH} in the analysis seasons

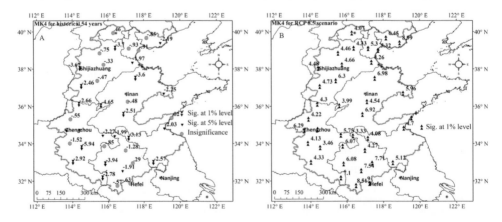

Chapter 2 Impacts of climate change on potential evapotranspiration under a historical period and future climate scenario... | 47

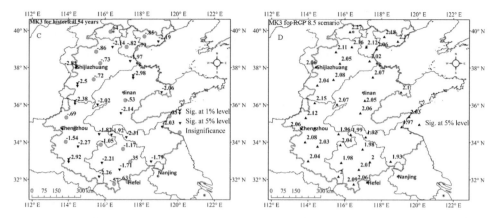

Figure 2-5 Z-statistic values of annual ET_0 and spatial pattern of ET_0 trends for the entire year using MK test considering all the significant autocorrelation structure (MK3) and MK test considering long-term persistence and Hurst coefficient (MK4) for historical data and RCP8.5 scenario. These maps were obtained with the software ArcGIS with down arrows for descending trend and upward arrows for ascending trend

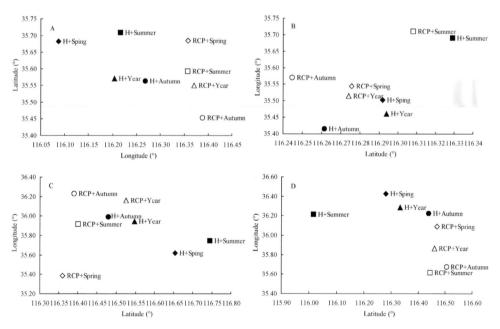

Figure 2-6 Sensitivity centers of ET_0 to (A) air temperature, (B) net radiation, (C) relative humidity and (D) wind speed during the historical period and under the RCP8.5 scenario. The coordinates of sensitivity centers for ET_0 to meteorological variables were estimated using barycentric fitting method from 40 stations during the historical period and under the RCP8.5 scenario. The barycentric fitting method comes from literatures reported by Xu et al. (2004), He et al. (2004) and Liu et al. (2012)

(except winter) for the past 54 years are all found in Shandong Province corresponding to the winter wheat and summer maize growing seasons (Yang et al., 2013). In comparison with the past 54 years, the centers of S_T and S_{WS} move eastward under the RCP8.5 scenario, whereas the centers of S_{RS} and S_{RH} move northwestward.

2.3.4 Regional response of ET_0 to climate change

The sensitivity of ET_0 to changes in meteorological factors was identified through sensitivity analysis during the four seasons and the entire year under the historical and future conditions; however, because of the variation and interaction of effect of the climatic variables, controlling factors to ET_0 remain difficult to pin down. Therefore, we used a second tool, a multivariate regression model, to get a relatively complete answer to this question: a multivariate regression model. The tendencies and magnitudes of the climate variables and the relationships between ET_0 and the key climatic variables in four seasons and entire year are presented in Table 2-5. The historical data set displays a decline of RH as the primary contributor to a negative trend in ET_0 in spring and over the entire year. The decline of RS is primarily responsible for a negative trend in ET_0 in summer, autumn, and winter. Different from the result under the historical condition, T_a becomes the primary controlling factor in the increment of ET_0 in spring under the RCP8.5 scenario. As for summer and the entire year, the increase of RS is responsible for the increment of ET_0, while the primary controlling variable turned to be RH on ET_0 in autumn and winter.

Table 2-5 Response of ET_0 to climate change during the historical period and under the RCP8.5 scenario

	T_a	RS	RH	WS
Historical				
Spring	0.34**	0.42**	0.77**	0.26**
Summer	0.45**	0.68**	0.52**	0.16**
Autumn	0.48**	0.63**	0.25**	0.10*
Winter	0.39**	0.94**	0.03	−0.05
Entire year	0.32**	0.56**	0.64**	0.23**
RCP8.5				
Spring	0.52**	0.42**	−0.37**	−0.24**
Summer	0.48**	0.50**	−0.35**	−0.29**
Autumn	0.59**	0.30**	−0.71**	−0.28**
Winter	0.25**	0.47**	−0.84**	−0.21**
Entire year	0.55**	0.58**	−0.39**	−0.48**

Notes: The values described in the table were obtained through a multivariate regression analysis between ET_0 and meteorological variables (air temperature, solar radiation, relative humidity, and wind speed) in the past 54 years and future 85 years over the Huang-Huai-Hai Plain

2.4 Discussion

2.4.1 Spatio-temporal evolution of ET_0

Under the effects of climate change, declining trends in ET_0 have been detected in various regions of the world (Jhajharia et al., 2012; Roderick and Farquhar, 2004; Thomas, 2000). Mcvicar et al. (2012) reported a global decline of evaporative demand at rate of 1.31 mm·y^{-1} by reviewing papers on ET_0 trend estimation ($n = 26$), whereas Thomas (2000) estimated that the rate in China was 2.3 mm·y^{-1} on the basis of a time series from 1954 to 1993. In our study, the annual ET_0 averaged over the entire Huang-Huai-Hai Plain revealed a general and significant decreasing trend of -1.29 mm·y^{-1}, which was higher than those over the Yellow River Basin (Liu et al., 2014; Ma et al., 2012), the Yangtze River catchment (Gong et al., 2006; Xu et al., 2006), and Haihe River Basin (Tang et al., 2011; Wang et al., 2010). such difference was mainly due to the greater decrement of RH over the entire Huang-Huai-Hai Plain (as the governing climatic variable for ET_0). However, Fan et al. (2016) found that the annual ET_0 at 37.5% of stations exhibited positive trends. These stations are mainly located in the mountainous plateau zone in China, and show an increasing ET_0 trend similar to those in northeastern China (Liang et al., 2010) and the Shiyang River Basin of northwest China (Su et al., 2015). The reason for this is climate warming, which turns out to be the most significant factor in an increase in annual ET_0 in these places. The declining magnitude of annual ET_0 over the Huang-Huai-Hai Plain was far lower than those of India (Jhajharia et al., 2015) and Iran (Dinpashoh et al., 2011) due to greater reduction of RH. This increment of ET_0 during spring is bad news for the wheat production in the region since precipitation is also scarce during the period and water resources are already over-exploited. Whereas ET_0 was slightly decreasing in the historical data set, ET_0 will increase in the following 74 years as predicted under the RCP8.5 scenario. Similar predictions are made for the Loess Plateau under A2 and B2 scenarios (Li et al., 2012). If this RCP scenario comes close to reality, the already elevated water demand will continue to increase, which poses serious questions to the sustainability of the current cropping practices in the Huang-Huai-Hai Plain. Accordingly, results have to be combined with information on the spatio-temporal distribution and trends of precipitation to identify when the water requirement of the agricultural crops cannot be met and to determine whether the groundwater resources are being used in an unsustainable way (Xu and Xu, 2012). It must also be noted that potential evapotranspiration is only a crude proxy for potential crop evapotranspiration since it does not take into account the specific water strategy of the crops, nor growth dynamics of them. Furthermore, change in drought condition is a consequence of climate change and aridity is usually expressed as a generalized function of precipitation, temperature, and/or ET_0 all reflecting the degree of meteorological drought. Over last decades, the aridity index (which is defined as the ratio of ET_0 to precipitation) is observed to have different trends in different regions (Liu et al., 2013b; Su et al., 2015; Wu et al., 2006). In our study, a declining trend in aridity was witnessed during the historical period and under the RCP8.5 scenario over the entire Huang-Huai-Hai Plain, which meant that the region started wetting and is thereby in accordance with the findings in North Iran (Ahani et al., 2013) and northwest China (Huo et al., 2013; Liu et al., 2013b).

2.4.2 Impact of meteorological variables on ET_0

Even though the variability across the plain is not very large, some areas do stand out with a significant increase of ET_0. If such area is not compensated by an increase in rainfall or other management practices, it could pose problems to the water allocation. Overall in the region, ET_0 is most influenced by relative humidity as also observed in the Wei River Basin (Zuo et al., 2012) and the Yangtze River Basin in China (Gong et al., 2006), followed by solar radiation and mean temperature. Goyal (2004) reported that the temperature increase contributed most to the ET_0 increase over an arid zone of Rajasthan in India. Yu et al. (2002) reported that the ET_0 increase in southern Taiwan of China was due to changes in temperature and RH in local paddy fields. Hess (1998) reported that increasing wind speed was responsible for the ET_0 increase in East Nigeria. While, decrease in wind speed, termed 'stilling' by Roderick et al. (2007) was identified to be responsible for the ET_0 decline in Australia (Roderick et al., 2007), the Tibetan Plateau (Chen et al., 2006) and the south of Canada (Burn and Hesch, 2007). These findings are inconsistent with our results that declining annual ET_0 was primarily resulting from the change in relative humidity in the Huang-Huai-Hai Plain. Indeed the PM equation dictates that the higher the RH, the lower the ET_0 should be. Nevertheless, our time series analysis indicated that the negative influence on ET_0 became weaker with time, which is in agreement with the result for Haihe River Basin and Huaihe River drainage system (Liu et al., 2012). Furthermore, the decline of relative humidity was primarily responsible for a negative trend in ET_0 in spring and the entire year. In contrast, ET_0s during spring and winter were most sensitive to changes in RS and WS, as also reported by Liu et al. (2011a) and Todisco and Vergni (2008) respectively. The decline of RS, also called global dimming, resulted in a declining trend of ET_0 in the Huang-Huai-Hai Plain, as estimated for many places (Stanhill and Cohen, 2001; Wild, 2014; Yang et al., 2014). A negative trend in ET_0 is controlled by the declining of RS in summer, autumn, and winter, as similarly found in the study of Yellow River Basin (Liu et al., 2014). When going from historical data to future conditions according to the RCP8.5 scenario, ET_0 remains most sensitive to changes in RH in autumn and over the entire year.

2.4.3 Estimated precipitation deficit and impact on agriculture

The fluctuation of ET_0 due to climate warming is expected to have important consequences because of its association with precipitation change. Zhai and Pan (2003) showed that annual precipitation has decreased by 30 mm in North China in the recent 50 years. In this study, a crude estimation of the precipitation deficit (PD) was established by subtracting the monthly cumulative ET_0 from the monthly cumulative precipitation (P). This approach has been used previously in agro-ecological studies of the Arabian Peninsula (De Pauw, 2002) and Northern China (Liu et al., 2013a). Positive and negative values indicate an excess or deficit in crop water requirements, respectively. It should be noted that PD was estimated using the ET_0 instead of the actual crop evapotranspiration (Harmsen et al., 2009), which makes it only a crude estimation. The annual variation tendency and statistics of the PD are described in Table 2-6 during the historical period and under the RCP8.5 scenario. It can be seen that a deficit in terms of crop water requirements is detected in almost all four seasons (except winter) in the past 54 years dataset. The same behavior was observed at the scale of the entire year for both scenarios, but PD is bound to become stronger under the RCP8.5 scenario. There are already seemingly increasing deficits in spring and winter, and negative consequences on wheat production and water resources in the area can be expected in the future if no management changes, such as irrigation techniques, cropping pattern, mulching, etc. are adopted (Allen et al., 1998; Olesen and Bindi, 2002). Based on precipitation data, ET_0 estimates and accurate crop coefficients adapted irrigation schedules could be defined, and strategies could be developed to optimize water use as well as yield. The present study gives the first hint of trends to be expected and regions to be prioritized. A study in North China revealed that mean daily soil evaporation with less and additional mulching was reduced by 16% and 37% for wheat production, respectively (Chen et al., 2007b). In Spain, Döll (2002) has predicted a decline in irrigation requirements by 2020 on account of the possibility of

Table 2-6 Trend analysis of precipitation deficit with Mann-Kendall test and Theil-Sen's slope estimator during the historical period and under the RCP8.5 scenario

Scenario	Spring		Summer		Autumn		Winter		Entire year	
	Value (mm)	Slope	Value (mm)	Slope	Value (mm)	Slope	Value (mm)	Slope	Value (mm)	Slope
Historical	−252.9	2.35	−187.6	−14.1**	31.32	−0.19	−82.8	2.58	−492.5	18.3**
RCP8.5	−221.1	−4.12*	18.9	−1.27	−166.7	−3.62	−65.5	−0.72*	−434.4	−9.73*

Notes: The units of the slope are mm·decade^{-1}. The values described in the table were obtained from area-averaged precipitation deficit from 40 meteorological stations using Theil-Sen's slope estimator in the past 54 years and future 85 years over the Huang-Huai-Hai Plain

earlier sowing under more favorable higher temperatures. It will be necessary to develop feasible straw (film) mulching, regulated deficit irrigation, and soil water storage and preservation to reduce pressure on groundwater over-exploitation, especially for winter wheat in the Huang-Huai-Hai Plain.

2.5 Conclusions

This study quantified the temporal evolution of meteorological variables and their influence on potential evapotranspiration as well as their spatial patterns over the Huang-Huai-Hai Plain, China using a historical data set of 54 years and a high CO_2 emission scenario RCP8.5 with daily time step. The historical data set revealed a decrease of ET_0 over time, which was only significant in summer and the entire year. Spring was the only season with a positive trend of ET_0. Conversely, ET_0 is predicted to increase for all seasons under the RCP8.5 scenario.

We quantified the sensitivity of ET_0 to the different key climate factors using a sensitivity analysis based on partial derivatives. The main controlling variable was not constant for all seasons. ET_0 was most sensitive to RH in summer and autumn, whereas RS and WS were more important during spring and winter. While a negative trend in ET_0 was primarily controlled by the decline of relative humidity in both the spring and the entire year. A negative trend in ET_0 was controlled by the decline of solar radiation in summer, autumn and winter. Also under the RCP8.5 scenario, RH remains a variable with a large influence on ET_0 during autumn and over the entire year. The impact of RS and WS is also in the same direction as in the historical data set, while the increase of RS is responsible for the increment of ET_0 in the entire year. The average temperature was the variable to which ET_0 was most sensitive in summer and its negative influence on ET_0 is predicted to strengthen in the future 74 years. Analysis of the spatial patterns revealed that ET_0 had been the lowest in the southeastern Shandong, and the highest in the northeastern part of the province. Overall no distinct spatial structure is visible in the Huang-Huai-Hai Plain in the historical data set, but the predictions under the RCP8.5 scenario result in a northeast-west gradient with low values in the south, mainly visible in spring and summer.

The accurate estimation of ET_0 response to meteorological variables could be favorable for designing future irrigation systems and agricultural production practices in the studied region. Variations in ET_0 in regards to meteorological parameters and agricultural production practice should be given greater attention when planning the hydrological process and water management in coming decades under the effects of global climate change, in order to avoid problems such as the over-exploitation of groundwater resources.

References

Ahani H, Kherad M, Kousari M R, Roosmalen L V, Aryanfar R, Hossenini S M. 2013. Non-parametric trend analysis of the aridity index for three large arid and semi-arid basins in Iran. Theoretical & Applied Climatology, 112(3-4): 553-564.

Allen R G, Pereira L S, Raes D, Smith M. 1998. Crop evapotranspiration: Guidelines for computing crop water requirements. FAO Irrigation and drainage paper 56. FAO, Rome, 300: 6541.

Brown R A, Rosenberg N J. 1997. Sensitivity of crop yield and water use to change in a range of climatic factors and CO_2 concentrations: A simulation study applying EPIC to the central USA. Agricultural and Forest Meteorology, 83(3): 171-203.

Burn D H, Hesch N M. 2007. Trends in evaporation for the Canadian Prairies. Journal of Hydrology, 336(1-2): 61-73.

Chen H, Guo S, Xu C, Singh V P. 2007a. Historical temporal trends of hydro-climatic variables and runoff response to climate variability and their relevance in water resource management in the Hanjiang basin. Journal of Hydrology, 344(3): 171-184.

Chen J, Tang C, Shen Y, Sakura Y, Kondoh A, Shimada J. 2003. Use of water balance calculation and tritium to examine the dropdown of groundwater table in the piedmont of the North China Plain (NCP). Environmental Geology, 44(5): 564-571.

Chen S, Liu Y, Thomas A. 2006. Climatic change on the Tibetan Plateau: Potential evapotranspiration trends from 1961–2000. Climatic Change, 76(3-4): 291-319.

Chen S, Zhang X, Pei D, Sun H, Chen S. 2007b. Effects of straw mulching on soil temperature, evaporation and yield of winter wheat: Field experiments on the North China Plain. Annals of Applied Biology, 150(3): 261-268.

Chun K, Wheater H, Onof C. 2012. Projecting and hindcasting potential evaporation for the UK between 1950 and 2099. Climatic Change, 113(3-4): 639-661.

Cohen S, Ianetz A, Stanhill G. 2002. Evaporative climate changes at Bet Dagan, Israel, 1964–1998. Agricultural and Forest Meteorology, 111(2): 83-91.

De Pauw E. 2002. An agroecological exploration of the Arabian Peninsula. Syria, ICARDA report.

Dinpashoh Y, Jhajharia D, Fakheri-Fard A, Singh V P, Kahya E. 2011. Trends in reference crop evapotranspiration over Iran. Journal of Hydrology, 399(3-4): 422-433.

Döll P. 2002. Impact of climate change and variability on irrigation requirements: A global perspective. Climatic Change, 54(3): 269-293.

Donohue R J, McVicar T R, Roderick M L. 2010. Assessing the ability of potential evaporation formulations to capture the dynamics in evaporative demand within a changing climate. Journal of Hydrology, 386(1): 186-197.

Drogue G, Pfister L, Leviandier T, Idrissi A E, Iffly J F, Matgen P, Humbert J, Hoffmann L. 2004. Simulating the spatio-temporal variability of streamflow response to climate change scenarios in a mesoscale basin. Journal of Hydrology, 293(1): 255-269.

Dyck S. 1985. Overview on the present status of the concepts of water balance models. IAHS-AISH Publication, 148: 3-19.

Fan J, Wu L, Zhang F, Xiang Y, Zheng J. 2016. Climate change effects on reference crop evapotranspiration across different climatic zones of China during 1956–2015. Journal of Hydrology, 542: 923-937.

Feddema J J. 1999. Future African water resources: Interactions between soil degradation and global warming. Climatic Change, 42(3): 561-596.

Gaile G L. 1984. Measures of spatial equality, Spatial statistics and models. New York: Springer International Publishing: 223-233.

Gao G, Chen D, Ren G, Chen Y, Liao Y. 2006. Spatial and temporal variations and controlling factors of potential evapotranspiration in China: 1956–2000. Journal of Geographical Sciences, 16(1): 3-12.

Georgia A. 2000. The change of the hydrological cycle under the influence of global warming. Hydrology for the Water Management of Large River Basis, 201: 119-128.

Gong L, Xu C, Chen D, Halldin S, Chen Y D. 2006. Sensitivity of the Penman-Monteith reference evapotranspiration to key climatic variables in the Changjiang (Yangtze River) basin. Journal of Hydrology, 329(3): 620-629.

Goyal R. 2004. Sensitivity of evapotranspiration to global warming: A case study of arid zone of Rajasthan (India). Agricultural Water Management, 69(1): 1-11.

Han S, Tang Q, Xu D, Wang S. 2014. Irrigation-induced changes in potential evaporation: More attention is needed. Hydrological Processes, 28(4): 2717-2720.

Harmsen E W, Miller N L, Schlegel N J, Gonzalez J E. 2009. Seasonal climate change impacts on evapotranspiration, precipitation deficit and crop yield in Puerto Rico. Agricultural Water Management, 96(7): 1085-1095.

He J, Wang Q J, Li H W, Liu L J, Gao H W. 2009. Effect of alternative tillage and residue cover on yield and water use efficiency in annual double cropping system in North China Plain. Soil and Tillage Research, 104(1): 198-205.

He Y, Zhang B, Ma C. 2004. Dynamic change of cultivated land and its impact on grain-production in Jilin Province. Resources Science, 26(4): 119-125. (in Chinese)

Hess T M. 1998. Trends in reference evapo-transpiration in the North East Arid Zone of Nigeria, 1961–1991. Journal of Arid Environments, 38(1): 99-115.

Hobbins M T, Ramírez J A, Brown T C. 2001. The complementary relationship in estimation of regional evapotranspiration: An enhanced advection-aridity model. Water Resources Research, 37(5): 1389-1403.

Huntington T G. 2006. Evidence for intensification of the global water cycle: Review and synthesis. Journal of Hydrology, 319(1): 83-95.

Huo Z, Dai X, Feng S, Kang S, Huang G. 2013. Effect of climate change on reference evapotranspiration and aridity index in arid region of China. Journal of Hydrology, 492(492): 24-34.

Irmak S, Payero J O, Martin D L, Irmak A, Howell T A. 2006. Sensitivity analyses and sensitivity coefficients of standardized daily ASCE-Penman-Monteith equation. Journal of Irrigation and Drainage Engineering, 132(6): 564-578.

Jhajharia D, Dinpashoh Y, Kahya E, Singh V P, Fakheri-Fard A. 2012. Trends in reference evapotranspiration in the humid region of northeast India. Hydrological Processes, 26(3): 421-435.

Jhajharia D, Kumar R, Dabal P P, Singh V P, Choudhary R R, Dinpashoh Y. 2015. Reference evapotranspiration under changing climate over the Thar Desert in India. Meteorological Applications, 22(3): 425-435.

Kendall M. 1975. Rank Correlation Methods. London: Charles Griffin & Company Ltd.

Kousari M R, Ahani H. 2012. An investigation on reference crop evapotranspiration trend from 1975 to 2005 in Iran. International Journal of Climatology, 32(15): 2387-2402.

Kumar S, Merwade V, Kam J, Thurner K. 2009. Streamflow trends in Indiana: Effects of long term persistence, precipitation and subsurface drains. Journal of Hydrology, 374(1-2): 171-183.

Li Z, Zheng F L, Liu W Z. 2012. Spatiotemporal characteristics of reference evapotranspiration during 1961–2009 and its projected changes during 2011–2099 on the Loess Plateau of China. Agricultural and Forest Meteorology, 154: 147-155.

Liang L, Li L, Liu Q. 2010. Temporal variation of reference evapotranspiration during 1961–2005 in the Taoer River basin of Northeast China. Agricultural and Forest Meteorology, 150(2): 298-306.

Liang W, Carberry P, Wang G, Lu R, Lu H, Xia A. 2011. Quantifying the yield gap in wheat-maize cropping systems of the Hebei Plain, China. Field Crops Research, 124(2): 180-185.

Liu B, Qi H, Wang W P, Zeng X F, Zhai J Q. 2011a. Variation of actual evapotranspiration and its impact on regional water resources in the Upper Reaches of the Yangtze River. Quaternary International, 244(2): 185-193.

Liu C, Zhang D, Liu X, Zhao C. 2012. Spatial and temporal change in the potential evapotranspiration sensitivity to meteorological factors in China (1960–2007). Journal of Geographical Sciences, 22(1): 3-14.

Liu Q, Mei X, Yan C, Ju H, Yang J. 2013a. Dynamic variation of water deficit of winter wheat and its possible climatic factors in Northern China. Acta Ecologica Sinica, 33(20): 6643-6651.

Liu Q, Yan C, Mei X, Zhang Y, Yang J, Liang Y. 2012. Spatial evolution of reference crop evapotranspiration in arid area of Northwest China. Chinese Journal of Agrometeorology, 33(1): 48-53. (in Chinese)

Liu Q, Yan C, Zhao C, Yang J, Zhen W. 2014. Changes of daily potential evapotranspiration and analysis of its sensitivity coefficients to key climatic variables in Yellow River basin. Transactions of Chinese Society of Agricultural Engineering, 30(15): 157-166.

Liu S, Mo X, Lin Z, Xu Y, Ji J, Wen G, Jeff R. 2010. Crop yield responses to climate change in the Huang-Huai-Hai Plain of China. Agricultural Water Management, 97(8): 1195-1209.

Liu S, Mo X, Lin Z, Xu, Y, Ji J, Wen G, Jeff R. 2010. Crop yield responses to climate change in the Huang-Huai-Hai-Plain of China. Agricultural Water Management, 97(8): 1195-1209.

Liu X, Zhang D, Luo Y, Liu C. 2013b. Spatial and temporal changes in aridity index in northwest China: 1960 to 2010. Theoretical & Applied Climatology, 112(1-2): 307-316.

Ma X, Zhang M, Li Y, Wang S, Ma Q, Liu W. 2012. Decreasing potential evapotranspiration in the Huanghe River Watershed in climate warming during 1960–2010. Journal of Geographical Sciences, 22(6): 977-988.

Mann H B. 1945. Nonparametric test against trend. Econometrica, 13(3): 245-259.

Maracchi G, Sirotenko O, Bindi M. 2005. Impacts of present and future climate variability on agriculture and forestry in the temperate regions: Europe. Climatic Change, 70(1-2): 117-135.

McKenney M S, Rosenberg N J. 1993. Sensitivity of some potential evapotranspiration estimation methods to climate change. Agricultural and Forest Meteorology, 64(1-2): 81-110.

McNulty S G, Vose J M, Swank W T. 1997. Regional hydrologic response of loblolly pine and air temperature and precipitation. Journal of the American Water Resources Association, 33(5): 1011-1022.

Mcvicar T R, Roderick M L, Donohue R J, Li L, Niel T G V, Thomas A, Grieser J, Jhajharia D, Himri Y, Mahowald N M. 2012. Global review and synthesis of trends in observed terrestrial near-surface wind speeds: Implications for evaporation. Journal of Hydrology, s416-417(3): 182-205.

Muhs D, Maat P. 1993. The potential response of eolian sands to greenhouse warming and precipitation reduction on the Great Plains of the USA. Journal of Arid Environments, 25(4): 351-361.

Nalley D, Adamowski J, Khalil B. 2012. Using discrete wavelet transforms to analyze trends in streamflow and precipitation in Quebec and Ontario (1954–2008). Journal of Hydrology, 475: 204-228.

Olesen J E, Bindi M. 2002. Consequences of climate change for European agricultural productivity, land use and policy. European Journal of Agronomy, 16(4): 239-262.

Rana G, Katerji N. 1998. A measurement based sensitivity analysis of the penman-monteith actual evapotranspiration model for crops of different height and in contrasting waster status. Theoretical and Applied Climatology, 60: 141-149.

Ren J, Chen Z, Zhou Q, Tang H. 2008. Regional yield estimation for winter wheat with MODIS-NDVI data in Shandong, China. International Journal of Applied Earth Observation and Geoinformation, 10(4): 403-413.

Riahi K, Rao S, Krey V, Cho C, Chirkow V, Fischer G, Kindermann G, Nakicenovic N, Rafaj P. 2011. RCP8.5: A scenario of comparatively high greenhouse gas emissions. Climatic Change, 109(1-2): 33-57.

Roderick M L, Farquhar G D. 2004. Changes in Australian pan evaporation from 1970 to 2002. International Journal of Climatology, 24(9): 1077-1090.

Roderick M L, Rotstayn L D, Farquhar G D, Hobbins M T. 2007. On the attribution of changing pan evaporation. Geophysical Research Letters, 34(34): 251-270.

Sen P K. 1968. Estimates of the Regression Coefficient Based on Kendall's Tau. Journal of the American Statistical Association, 63(324): 1379-1389.

Song F, Zhou T, Qian Y. 2014. Response of East Asian summer monsoon to natural and anthropogenic forcings in the 17 latest CMIP5 models. Geophysical Research Letters, 41: 596-603.

Song Z, Zhang H, Snyder R, Anderson F, Chen F. 2009. Distribution and trends in reference evapotranspiration in the North China Plain. Journal of Irrigation and Drainage Engineering, 136(4): 240-247.

Stanhill G, Cohen S. 2001. Global dimming: A review of the evidence for a widespread and significant reduction in global radiation with discussion of its probable causes and possible agricultural consequences. Agricultural and forest meteorology, 107(4): 255-278.

Stocker T, Qin D, Plattner G K, Tignor M, Allen S K, Boschung J, Nauels A, Xia Y, Bex V, Midgley P M. 2014. Climate Change 2013: The physical science basis. Cambridge, UK, and New York: Cambridge University Press.

Su X, Singh V P, Niu J, Hao L. 2015. Spatiotemporal trends of aridity index in Shiyang River basin of northwest China. Stochastic Environmental Research and Risk Assessment, 29(6): 1571-1582.

Suat I, Kabenge I, Skaggs KE, Mutiibwa D. 2012. Trend and magnitude of changes in climate variables and reference evapotranspiration over 116-yr period in the Platte River Basin, central Nebraska-USA. Journal of Hydrology, 420-421: 228-244.

Sun Q, Krobel R, Muller T, Romheld V, Cui Z, Zhang F, Chen X. 2011. Optimization of yield and water-use of different cropping systems for sustainable groundwater use in North China Plain. Agricultural Water Management, 98(5): 808-814.

Tang B, Tong L, Kang S Z, Zhang L. 2011. Impacts of climate variability on reference evapotranspiration over 58 years in the Haihe river basin of north China. Agricultural Water Management, 98(10): 1660-1670.

Team R D C. 2011. R: A Language and Environment for Statistical Computing. Computing, 14: 12-21.
Theil H. 1992. A Rank Invariant Method of Linear and Polynomial Regression Analysis. Netherlands: Springer: 345-381.
Thomas A. 2000. Spatial and temporal characteristics of potential evapotranspiration trends over China. International Journal of Climatology, 20(4): 381-396.
Thomas A. 2008. Agricultural irrigation demand under present and future climate scenarios in China. Global and Planetary Change, 60(3): 306-326.
Todisco F, Vergni L. 2008. Climatic changes in Central Italy and their potential effects on corn water consumption. Agricultural and Forest Meteorology, 148(1): 1-11.
Wang W, Peng S, Yang T, Shao Q, Xu J, Xing W. 2010. Spatial and temporal characteristics of reference evapotranspiration trends in the Haihe River basin, China. Journal of Hydrologic Engineering, 16(3): 239-252.
Wang Y, Jiang T, Bothe O, Fraedrich K. 2007. Changes of pan evaporation and reference evapotranspiration in the Yangtze River basin. Theoretical and Applied Climatology, 90(1-2): 13-23.
Wild M. 2014. Global dimming and brightening. In: Bill F. Global Environmental Change. Berlin: Springer: 39-47.
Wu S, Yin Y, Zheng D, Yang Q. 2006. Moisture conditions and climate trends in China during the period 1971–2000. International Journal of Climatology, 26(2): 193-206.
Xu C, Gong L, Jiang T, Chen D, Singh V P. 2006. Analysis of spatial distribution and temporal trend of reference evapotranspiration and pan evaporation in Changjiang (Yangtze River) catchment. Journal of Hydrology, 327(1-2): 81-93.
Xu C, Singh V. 2005. Evaluation of three complementary relationship evapotranspiration models by water balance approach to estimate actual regional evapotranspiration in different climatic regions. Journal of Hydrology, 308(1): 105-121.
Xu C, Xu Y. 2012. The projection of temperature and precipitation over China under RCP scenarios using a CMIP5 multi-model ensemble. Atmospheric and Oceanic Science Letters, 5(6): 527-533.
Xu X, Liu J, Zhuang D. 2004. Analysis of temporal-spatial characteristics of reference evapotranspiration based on GIS technology in Northeast China during 1991–2000. Transactions of the CSAE, 20(2): 10-14. (in Chinese)
Xu Z X, Takeuchi K, Ishidaira H. 2003. Monotonic trend and step changes in Japanese precipitation. Journal of Hydrology, 279(1-4): 144-150.
Xu Z, Li J. 2003. A distributed approach for estimating catchment evapotranspiration: Comparison of the combination equation and the complementary relationship approaches. Hydrological Processes, 17(8): 1509-1523.
Yang J, Liu Q, Mei X, Yan C, Ju H, Xu J. 2013. Spatiotemporal characteristics of reference evapotranspiration and its sensitivity coefficients to climate factors in Huang-Huai-Hai Plain, China. Journal of Integrative Agriculture, 12(12): 2280-2291.
Yang J, Liu Q, Yan C, Mei X. 2011. Spatial and temporal variation of solar radiation in recent 48 years in North China. Acta Ecologica Sinica, 31(10): 2748-2756.
Yang J, Mei X, Huo Z, Yan C, Ju H, Zhao F, Liu Q. 2015. Water consumption in winter wheat and summer maize cropping system based on SEBAL Model in Huang-Huai-Hai Plain, China. Journal of Integrative Agriculture, 14(10): 2065-2076.
Yang K, Wu H, Qin J, Lin C, Tang W, Chen Y. 2014. Recent climate changes over the Tibetan

Plateau and their impacts on energy and water cycle: A review. Global and Planetary Change, 112: 79-91.

Yu P S, Yang T C, Chou C C. 2002. Effects of climate change on evapotranspiration from paddy fields in Southern Taiwan. Climatic Change, 54(1-2): 165-179.

Yuan B, Guo J, Ye M, Zhao J. 2012. Variety distribution pattern and climatic potential productivity of spring maize in Northeast China under climate change. Chinese Science Bulletin, 57(26): 3497-3508.

Yue S, Pilon P, Phinney B, Cavadias G. 2002. The influence of autocorrelation on the ability to detect trend in hydrological series. Hydrological Processes, 16(9): 1807-1829.

Zamani R, Mirabbasi R, Abdollahi S, Jhajharia D. 2016. Streamflow trend analysis by considering autocorrelation structure, long-term persistence, and Hurst coefficient in a semi-arid region of Iran. Theoretical and Applied Climatology, 129(1-2): 33-45.

Zeng W, Heilman J. 1997. Sensitivity of evapotranspiration of cotton and sorghum in west Texas to changes in climate and CO_2. Theoretical and Applied Climatology, 57(3-4): 245-254.

Zhai P, Pan X. 2003. Change in extreme temperature and precipitation over Northern China during the second half of the 20th century. Acta Ecologica Sinica, 58(S1): 2-10.

Zhang Q, Xu C, Chen X. 2011. Reference evapotranspiration changes in China: Natural processes or human influences. Theoretical and Applied Climatology, 103(3-4): 479-488.

Zhang X, Kang S, Zhang L, Liu J. 2010. Spatial variation of climatology monthly crop reference evapotranspiration and sensitivity coefficients in Shiyang river basin of northwest China. Agricultural Water Management, 97(10): 1506-1516.

Zhao R, Chen X, Zhang F, Zhang H, Schroder J, Romheld V. 2006. Fertilization and nitrogen balance in a wheat-maize rotation system in North China. Agronomy Journal, 98(4): 938-945.

Zheng C, Wang Q. 2014. Spatiotemporal variations of reference evapotranspiration in recent five decades in the arid land of Northwestern China. Hydrological Processes, 28(25): 6124-6134.

Zhou T, Hong T. 2013. Projected changes of palmer drought severity index under an RCP8.5 scenario. Atmospheric and Oceanic Science Letters, 6(5): 273-278.

Zuo D, Xu Z, Yang H, Liu X. 2012. Spatiotemporal variations and abrupt changes of potential evapotranspiration and its sensitivity to key meteorological variables in the Wei River basin, China. Hydrological Processes, 26(8): 1149-1160.

Chapter 3 Spatio-temporal variation of drought characteristics in the Huang-Huai-Hai Plain, China under the climate change scenario

Abstract

Understanding the potential drought characteristics under climate change is essential to reducing vulnerability and establish adaptation strategies, especially in the Huang-Huai-Hai Plain, which is the grain production base in China. In this paper, we investigated the variations in drought characteristics (drought event frequency, duration, severity, and intensity) for the past 50 years (1961–2010) and under future scenarios (2010–2099), based on the observed meteorological data and the RCP8.5 projection, respectively. First, we compared the applicability of three climatic drought indices: the standardized precipitation index (SPI), the standardized precipitation evapotranspiration index (SPEI) based on the Penman-Monteith equation (SPEI-PM), and the Thornthwaite equation (SPEI-TH) to identify the historical agricultural drought areas. Then, we analyzed the drought characteristics using "run theory" for both historical observations and future RCP8.5 scenarios based on proper index. Correlation analyses between drought indices and agricultural drought areas showed that SPEI-PM performed better than SPI and SPEI-TH in the Huang-Huai-Hai Plain. Based on the results of SPEI-PM, in the past 50 years the Huang-Huai-Hai plain has experienced a reduced drought of shorter duration, and weaker severity and intensity. However, under the future RCP8.5 scenario, drought is expected to rise in frequency, duration, severity, and intensity from 2010–2099, although drought components during the 2010–2039 is predicted to be milder compared with the historical conditions. This study highlights that the estimations for atmospheric evaporative demand would bring in differences in the prediction of long-term drought trends by different drought indices. The results of this study can help inform researchers and local policy makers to establish drought management strategies.

3.1 Introduction

Drought is one of the most damaging and widespread climate extremes that negatively affect the agricultural production, water resources, ecosystem function, and human lives around the world (Dai, 2011b; Wilhite et al., 2007). Due to the interaction of monsoon climate with the complex topography in East Asia, China has suffered from long-lasting and severe droughts during the second half of twentieth century, which has caused significant socioeconomic and eco-environmental damages to the country (Yang et al., 2015; Yong et al., 2013; Zhang et al., 2015a). With the projected temperature increase and change in the distribution of precipitation, the drought risk is expected to increase further (Sillmann et al., 2013; Wang and Chen, 2014) and subsequently make crop production more uncertain. Thus, understanding the potential drought characteristics under climate change is of prime importance for reducing vulnerability and establishing drought adaptation strategies (Chen et al., 2014; Wilhite et al., 2014).

Several techniques have been developed to quantitatively analyze drought characteristics (Heim, 2002). Some of these are the physically based indices, such as Palmer drought severity index (PDSI) and its derivative, and statistically based indices, such as standardized precipitation index (SPI) and the standardized precipitation evapotranspiration index (SPEI)(Vicente-Serrano et al., 2011). These indices have been widely used in detecting long-term drought trends under climate change at several locations around the world. The general recognition is that drought has been intensifying around the world due to global warming in the past decade s(Allen et al., 2010; Dai, 2013). However, it has been difficult to understand how droughts have changed in China, because the findings based on the potential evapotranspiration (ET_0) equations vary among studies (Sheffield et al., 2012; Trenberth et al., 2014; Xu et al., 2015). For example, significant drying trends were found in northern and southwestern China during the past decades, when ET_0 was estimated by temperature only (Wang et al., 2015a; Yu et al., 2014a). However, when it was calculated by the Penman-Monteith equation, whose algorithm takes more climatic variables into account, no evidence of an increase in drought severity was found across China (Wang et al., 2015b), and even more wetting areas than drying areas were observed in the North China Plain and Northeast China Plain (Xu et al., 2015). Subsequently, such differences have led to confusion among scientists, policy makers, and the public (Vicente-Serrano et al.,

2012). Thus, the applicability of an index should be verified before the implementation, in order to obtain results that are more realistic. Substantial divergence of applicability exists between physically and statistically based drought indices, and difference exists even when using the same index with different ET_0 estimation methods, depending on the system and location. Dai (2011a) reported that PDSI performed better than other indices because of its physically based water balance model, but Vicente-Serrano et al. (2011) described the advantages of statistically based indices, including SPI and SPEI and provided a worldwide rating of the performance of PDSI, SPI, and SPEI on hydrological, agricultural, and ecological systems (Vicente-Serrano et al., 2012). Similar comparative studies have been conducted at different locations for different systems, such as river discharges (Zhai et al., 2010), tree-ring growth (Sun and Liu, 2013), and crop production (Xu et al., 2013), resulting in site or system specific results.

The Huang-Huai-Hai Plain is a major crop producing area in China that encompasses 19% of the total arable land in China. However, it has experienced serious drought and water scarcity problems in recent years (Yong et al., 2013), which has been the limiting factor in agricultural production (Zhang et al., 2015a). Furthermore, water limitations are likely to be accentuated by increasing food demand, soil quality deterioration and over-exploitation of groundwater resources (Yang et al., 2015). Climate variability, especially extreme climatic events, such as drought, may cause fluctuation of crop yields (Lu and Fan, 2013; Yu et al., 2014b). Thus, understanding the potential variations of drought characteristics under climate change is essential for reducing vulnerability and establishing drought adaptation strategies for agriculture in the Huang-Huai-Hai Plain. Most previous studies have primarily reported the seasonal and spatial variability of water deficiency (Huang et al., 2014; Yong et al., 2013) and the long term drought evolutions, including dry/wet trends, spatial distribution of drought frequency, drought affected areas, and drought duration for historical periods (Wang et al., 2015a; 2015b; Xu et al., 2015; Yu et al., 2014a). However, few studies have evaluated the performance of multi-indices on estimating drought impact and assessed drought risk for future climate scenarios.

The objectives of this research are: (1) to evaluate the applicability of drought indices (SPI and two versions of SPEI) to accurate identification of historical agricultural drought areas in the Huang-Huai-Hai Plain; (2) to investigate the variations of drought characteristics, including duration, severity, and intensity in 1961–2010 and (3) to evaluate projected drought characteristics in 2010–2099 using the RCP8.5 climate scenario. The results are expected to provide useful information on drought risk to decision makers and a wide range of stakeholders interested in the occurrence and consequences of recurrent droughts.

3.2 Materials and methods

3.2.1 Study region

The Huang-Huai-Hai Plain is located in Northern China, extending over 31°14′–40°25′N and 112°33′–120°17′E, with 23.3 million ha of arable land (19% of the total arable land in China), providing about 70% of national wheat production and 30% of national maize production with a dominant winter wheat-summer maize double cropping system (Yang et al., 2015). The Huang-Huai-Hai Plain belongs to the extratropical monsoon climatic region. The annual mean precipitation is 500–800 mm with more than 70% rainfall in July to September and the atmospheric evaporative demand is about 1000 mm·y^{-1} (Zhang et al., 2011). For the wheat-maize rotation system, rainfall can only meet 65% of total agricultural water demand, and for winter wheat, only 25%–40% of water demand is satisfied by rainfall (Mei et al., 2013). The irrigation water is primarily pumped from groundwater. However, the groundwater level has decreased from a depth of 10 m in the 1970s to 32 m in 2001, and has continued to decrease at the rate of 1 m per year (Zhang et al., 2005). Thus, drought in this region not only challenges food supplies but also results in a series of environmental problems. The Huang-Huai-Hai Plain can be divided into six sub-regions (see Li et al., 2015 for detail information) in terms of climate conditions and agricultural management practices (Table 3-1).

Table 3-1 Drought classifications based on standardized precipitation index (SPI) and standardized precipitation-evapotranspiration index (SPEI)

Drought classes	Probability (%)	Index value
Extreme wet	2.3	⩾2.0
Very wet	4.4	1.5–2.0
Moderate wet	9.2	1.0–1.5
Near normal	68.2	−1.0–1.0
Moderate dry	9.2	−1.5–−1.0
Severe dry	4.4	−2.0–−1.5
Extreme dry	2.3	⩽−2.0

3.2.2 Climate data

Historical data covering a period of 50 years (1961–2010) and the interpolated future drought characteristics for the next 90 years (2010–2099) were used in this study. Historical meteorological data (maximum & minimum temperatures, wind speed, relative humidity, and daily sunshine duration) from 45 weather stations (Table 2-1) were obtained from the China Meteorological Administration (CMA). Future climate data were simulated using the RCP8.5 emission scenario of the HadGEM2-ES climate model, which assumes that the greenhouse gas emission continues to increase at the present rate. This dataset has been used for assessing potential effects of drought on winter wheat yield (Leng et al., 2015) and its vulnerability and adaptive capacity in the Huang-Huai-Hai Plain (Li et al., 2015).

3.2.3 Drought area data

Several previous studies have used crop yields data to verify the regional performance of drought indices (Ming et al., 2015; Potopova et al., 2015; Vicente-Serrano et al., 2012; Zhang et al., 2015b). However, in order to evaluate yield variability due to climate fluctuations alone, yield data should be detrended to remove the effect of agricultural technology improvements, such as better fertilizer application, new crop varieties, and better tillage practices (Potopova et al., 2015; Yu et al., 2014b). Similarly, it is difficult to determine how much yield is affected by drought. Thus, in this study, we used drought area data to verify the performance of SPI, SPEI-TH, and SPEI-PM.

The drought area data, including the Drought-Induced Areas (DIA), Drought-Affected Areas (DAA), and Lost Harvest Areas (LHA), were obtained from the disaster database of Department of Plantation, Ministry of Agriculture (http://www.zzys.moa.gov.cn/). The DIA, DAA, and LHA represent the arable areas with yield losses caused according to the recorded data by drought at 10%, 30%, and 70%, respectively. Thus, they actually reflect the cumulative effect of drought on harvest yield for the whole growing season of winter wheat and summer maize (the agricultural pattern is winter wheat and summer maize rotation system in the Huang-Huai-Hai Plain), and not the status of the conventional 'drought area' that measures the cover range of a certain drought event. The disaster database embodies provincial scale disaster data since 1949. However, considering the data's integrity and usability, data of DIA from 1971–2013, DAA from 1970–2012, LHA from 1982–2013 of Henan, Hebei, and Shandong provinces were considered only.

3.2.4 Calculations of drought indices

The performances of SPI and the two versions of SPEI in characterization of historical agricultural drought areas were evaluated. The SPI (McKee et al., 1993) is the number of standard deviations of the standardized normal distribution transformed from the precipitation (P) series, while the calculation of SPEI (Vicente-Serrano et al., 2010) follows the same procedure but on the basis of the precipitation (P) and the potential evapotranspiration (ET_0) series. To compute these indices, the probability density function of P sums series or P minus ET_0 sums series for a desired scale was estimated, which was then transformed to a normal distribution. As a result, the value of SPI or SPEI can be constructed as a split line that separates the standard normal distribution. The main strength of SPI and SPEI is that they can be calculated for any timescale (Heim, 2002), and represent the cumulative impact of drought for different time periods (Hayes et al., 2011). The SPI depends only on precipitation data, which makes it popular and easy to implement around the world. The SPEI takes into account the atmospheric evaporation demand (i.e., ET_0), which makes it suitable for drought analysis under climate change (Vicente-Serrano et al., 2012). In the original version of SPEI, the evapotranspiration was estimated by the Thornthwaite equation, in which only temperature is considered. However, the Penman-Monteith method is widely accepted as the most robust and accurate estimation of ET_0 as it includes the effects of temperature, wind speed, relative humidity, and solar radiation (Chen et al., 2005; Sentelhas et al., 2010). Thus, we calculated the SPI and SPEI using the Thornthwaite (SPEI-TH) and FAO-PM equations (SPEI-PM) to compare their correlation with the observed agricultural drought areas.

The SPI and two versions of SPEI (SPEI-TH and SPEI-PM) were calculated utilizing the R package of SPEI, developed by Begueria et al. (2014). Since SPI and SPEI are both standardized, their values should have the same statistical meaning, and therefore should be comparable. Thus, the same threshold as shown in Table 3-1 was used to classify the drought conditions.

3.2.5 Drought identification using run theory

The 'run theory' (Figure 3-1) proposed by Yevjevich (1967), which has been applied frequently to time series of anomalous hydrologic events, was used to identify drought components and investigate their statistical properties (Mishra and Desai, 2005; Nam et al., 2015). A run or a drought event is defined as a consecutive sequence of months (t) with drought indices values, X_t, less than a chosen threshold, X_0. A drought event is characterized by the following components, which can be used for mathematical analysis of drought. Drought initiation time (T_s) is the onset month of a drought event. Drought termination time (T_e) represents the date when the water shortage is relieved so that the drought no longer persists. Drought duration (D) is the time period between the initiation and termination of a drought. Drought severity (S) is obtained by the cumulative deficiency of the drought parameter below the critical level. Drought intensity (I) is calculated as the ratio of the drought deficit volume to the drought duration.

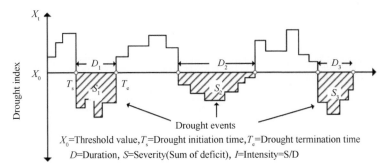

Figure 3-1 Drought characteristics identification using the run theory (According to Yevjevich, 1967). In our study, three drought indices were selected followed by standardized precipitation index (SPI) and two versions of the standardized precipitation evapotranspiration index (SPEI), SPEI-TH and SPEI-PM

3.3 Results

3.3.1 Selection of preferable drought index

As shown in Figure 3-2, drought areas have declined since the 1970s. Particularly, the DIAs of all the provinces have decreased significantly ($P<0.001$). However, the DAAs of the three provinces show insignificant declines, but remained at lower levels after 2003. The LHA of Henan (Figure 3-2-A3) and Shandong (Figure 3-2-C3) decreased significantly ($P<0.05$), while it was not significant for the Hebei province (Figure 3-2-B3).

The Pearson correlation coefficients (r) between monthly drought index series at 1–12 timescales and the agricultural drought areas in Hebei (HB), Henan (HN), and Shandong (SD) provinces, based on data from the Ministry of Agriculture for historical years, are summarized in Figure 3-3. It shows that the absolute value of Pearson's r increased from January to December. Thus, a 12-month SPEI-PM at the end of December in the Huang-Huai-Hai Plain could represent the overall yearly dry conditions that would lead to a crop yield reduction. Table 3-2 compares the Pearson's r between SPI, SPEI-PM, and SPEI-TH in December at 12-month scale. The SPEI-PM gave higher Pearson's r values than the SPI and SPEI-TH. Thus, SPEI-PM was considered to be a proper index for drought analysis in the Huang-Huai-Hai Plain.

3.3.2 Drought characteristics over the past 50 years

Historical drought evolution

The drought/wet evolutions computed by SPEI-PM at 1–24 month scales are depicted for the six sub-regions of the Huang-Huai-Hai Plain during 1961–2010 (Figure 3-4). The

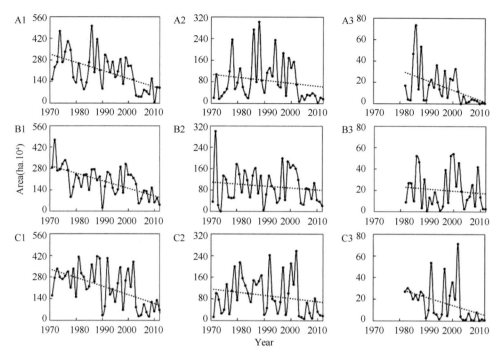

Figure 3-2 The series of drought area of Henan, Hebei, and Shandong Province. (A) Henan province; (B) Hebei province; (C) Shandong province. 1. drought-induced areas (DIA); 2. drought-affected areas (DAA); 3. lost harvest areas (LHA). The dotted line is the linear trend line

Table 3-2 Comparison of Pearson's *r* between SPEI-PM, SPEI-TH, and SPI at 12-month scale. Pearson correlation method comes from the literature reported by Ahlgren et al. (2003)

Province	Classification	DIA	DAA	LHA
HB	SPEI-PM	−0.54	−0.73	−0.74
	SPEI-TH	−0.26	−0.65	−0.58
	SPI	−0.44	−0.73	−0.60
HN	SPEI-PM	−0.48	−0.64	−0.58
	SPEI-TH	−0.29	−0.58	−0.45
	SPI	−0.44	−0.64	−0.56
SD	SPEI-PM	−0.75	−0.81	−0.78
	SPEI-TH	−0.57	−0.70	−0.71
	SPI	−0.68	−0.76	−0.74

Notes: DIA. drought-induced areas; DAA. drought-affected areas; LHA. lost harvest areas; HB. Hebei province; HN. Henan province; SD. Shandong province

horizontal axis represents month series from January 1961 to December 2010, while the vertical axis represents the timescales from 1–24. By utilizing the single diagram, the temporal trends of the severity and duration of the drought indices and the development

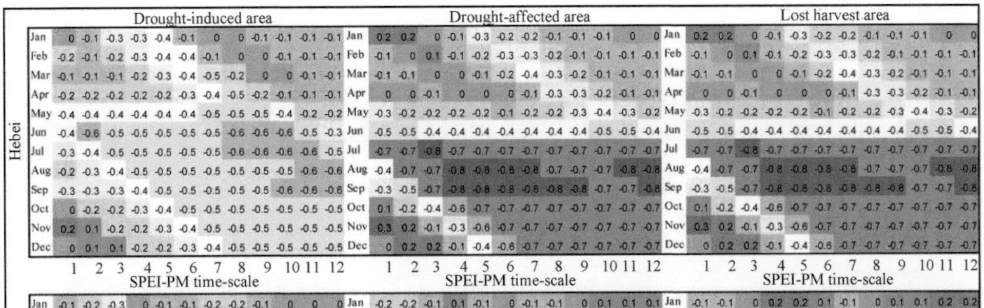

Figure 3-3 Pearson correlation coefficients between the monthly SPEI-PM series and the drought-induced areas, drought-affected areas, and lost harvest areas at 1–12 month scales. Pearson correlation method comes from the literature reported by Ahlgren et al. (2003)

of the drought/wet stress conditions from 1 to 24 timescales over the past 50 years can be easily identified. Furthermore, the characteristics of drought occurrences in different regions can be compared. According to Figure 3-4, the moderate ($\leqslant -1.0$) drought appeared in 1965–1970, 1980–1985, and 2000–2005 for all sub-regions. However, the severe droughts ($\leqslant -1.5$) occurred primarily before the 1970s. Furthermore, after the severe to extreme wet period around 1965, moderate-to-severe dry conditions occurred in all regions from 1965 to 1970. Particularly, sub-regions five and six (Figure 3-4-E and Figure 3-4-F) experienced extreme dry conditions during this period.

Temporal variation of historical drought characteristics

The changes in the number of total drought events were observed by 45 meteorological stations at multiple timescales in every decade of the historical periods (Table 3-3). A drought event was counted according to the 'run theory' (Figure 3-5). The threshold

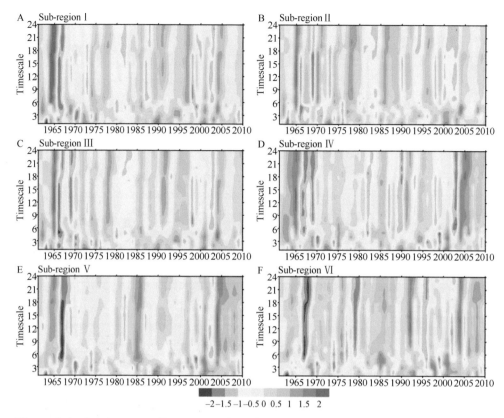

Figure 3-4 Spatio-temporal evolution of the SPEI-PM series indicating the development of drought from 1 to 24-month scales at 6 sub-regions during 1961–2010. These maps were obtained with the Surfer software

Table 3-3 Temporal changes in average drought intensity for all drought events in 1961–2010

Timescale	1960s	1970s	1980s	1990s	2000s
1-month	0.47	0.43	0.34	0.41	0.36
3-month	0.45	0.36	0.3	0.35	0.37
6-month	0.48	0.37	0.28	0.36	0.3
12-month	0.53	0.29	0.23	0.28	0.27
24-month	0.4	0.26	0.22	0.27	0.27

values are shown in Table 3-4. Drought events were categorized into two types: above-moderate drought events (hereinafter AM event) whose threshold was −1.0 and above-severe drought events (hereinafter AS event) whose threshold was −1.5. The ratio of AS events to AM events was also calculated to indicate changes in drought severity. An AM event includes all kinds of drought while an AS event includes more severe drought. In AM events, no significant trend was detected from 1960s to 2000s for all timescales (Table 3-4). AM events occurred infrequently in the 1960s and 2000s. However, the AS/AM ratio in the 1960s was found to be the highest with a gradual decrease toward the 2000s for different timescales. In the case of SPEI-PM for 6-month scale, the ratio decreased from 72.8% in the 1960s, to 51.0% in the 1970s, to 37.8% in the 1980s, and then to 35.5% in the 2000s. Thus, total drought events showed no significant tendency over the historical period, but drought events with more severity decreased gradually from the 1960s to 2000s.

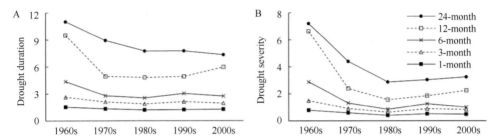

Figure 3-5 Temporal changes in average drought duration (A) and drought severity (B) for all drought events during 1961–2010

Table 3-4 Temporal variations in the number of drought events

Classification		1960s	1970s	1980s	1990s	2000s
1-month	AM event	610	741	628	717	560
	AS event	344(56.4)	298(40.2)	159(25.3)	270(37.7)	194(34.6)
3-month	AM event	380	456	425	407	361
	AS event	227(59.7)	202(44.3)	148(34.8)	189(46.4)	168(46.5)
6-month	AM event	224	296	328	315	262
	AS event	163(72.8)	151(51.0)	124(37.8)	165(52.4)	93(35.5)
12-month	AM event	122	142	188	187	148
	AS event	97(79.5)	65(45.8)	71(37.8)	102(54.5)	67(45.3)
24-month	AM event	77	108	151	101	111
	AS event	63(81.8)	48(44.4)	73(48.3)	50(49.5)	59(53.2)

Notes: AM events include all dry degree events, while AS events are for dry events more than the severe degree. Numbers in parentheses are the ratio of AS events to AM events in percentage

Drought event changes might not bring about the changes of other drought characteristics due to the association between drought duration, severity, and intensity. The average drought duration for almost all timescales decreased with fluctuations from the 1960s to the 2000s (Figure 3-5-A). Compared with the 1960s, the average drought duration in the 2000s decreased by 16.7% for the 1-month timescale, 25.7% for the 3-month timescale, 37.0% for the 6-month timescale, 37.2% for the 12-month timescale, and 33.5% for the 24-month timescale. Drought severity also decreased from the 1960s to 2000s (Figure 3-5-B). Compared with the 1960s, average drought severity in the 2000s decreased by 38.8% for the 1-month timescale, 44.5% for the 3-month timescale, 64.8% for the 6-month timescale, 66.2% for the 12-month timescale, and 54.9% for the 24-month timescale.

The average drought intensity for all drought events by decade shows decreased drought intensity from the 1960s to 2000s (Table 3-4). For all timescales, drought intensity was highest in the 1960s and lowest in the 1980s during the past 50 years. Compared with the 1960s, the drought intensity in the last decade was reduced by 23.4% for the 1-month timescale, 17.8% for the 3-month timescale, 37.5% for the 6-month timescale, 49.1% for the 12-month timescale, and 32.5% for the 24-month timescale.

3.3.3 Drought prediction for 2010–2099 under RCP8.5 scenario

Drought evolution under future climate change

The drought evolution maps of multi-scale SPEI-PM are shown in Figure 3-6 for six sub-regions during 2010–2099. Visual comparison of maps showed a tendency toward greater drought conditions through 2100. In the period (2010–2040), particularly in the second half of 2020–2030, all regions were mainly characterized by longer wet events. The severe dry event (SPEI-PM < –1.5) during this period can only be found during 2030s with a short dry event in sub-region I –IV. In the period (i.e. 2040–2070), except for sub-region VI and the end of 2060–2070, longer drought event is predicted to occur frequently, particularly during 2050s. In the period (i.e. 2070–2100), drought event is expected to occur persistently with higher frequency and longer duration during 2070s for region IV to VI and the end of 2080s for all regions. The main wet years were projected in the middle of 2070s for sub-regions I –III and the last ten years for sub-regions I –V.

Temporal variation of future drought characteristics

Changes in the drought events were compared between the three 30-year periods centered on 2025, 2055, and 2085 under the RCP8.5 climate scenario and the recent (1981–2010) historical period (Table 3-5). Both AM and AS events were predicted to be lower in the mid-period of 2010–2040 than in the historical period (1981–2010), indicating a lower drought frequency in the first 30 years in the future. In terms of SPEI-PM3, the drought frequency in the mid-period of 2010–2040 was projected to drop by 30.3% for AM events and 55.0% for AS events

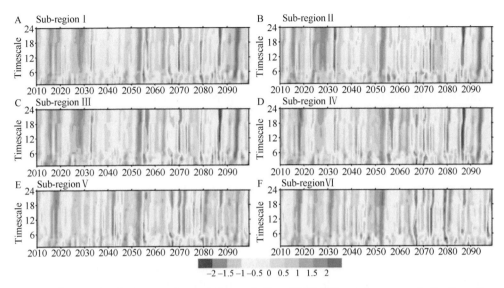

Figure 3-6 Spatio-temporal evolution of the SPEI-PM series was indicating the development of drought from 1 to 24 month-scales at six sub-regions under the RCP8.5 scenario. These maps were obtained with the Surfer software

Table 3-5 Drought event change under future weather scenario using the RCP8.5 climate scenario

Classification		1981–2010	RCP8.5		
			mid-period of 2010–2040	mid-period of 2040–2070	mid-period of 2070–2100
SPEI-PM1	AM event	1905	1321	2170	2242
	AS event	623(32.7)	360(27.3)	941(43.4)	1183(52.8)
SPEI-PM3	AM event	1193	832	1371	1430
	AS event	505(42.3)	227(27.3)	645(47.0)	825(57.7)
SPEI-PM6	AM event	905	605	981	1015
	AS event	382(42.2)	136(22.5)	483(49.2)	596(58.7)
SPEI-PM12	AM event	523	317	477	530
	AS event	240(45.9)	72(22.7)	251(52.6)	338(63.8)
SPEI-PM24	AM event	363	217	340	464
	AS event	182(50.1)	47(21.7)	198(58.2)	334(72.0)

Notes: AM events represent all dry events while AS events are for severe dry events. Numbers in parentheses are the ratio of AS events to AM events in percentage

lower compared with the historical standard. However, the frequency of drought events was predicted to continuously increase in the mid-period of 2040–2070 and mid-period of 2070–2100 for all time scales and surpasses that in the 1981–2010 period. In the case of the SPEI-PM6, compared with the historical data

(1981–2010), the results indicated an increase in moderate dry events in the mid-period of 2040–2070 and the mid-period of 2070–2100 at the rate of 8.4% and 12.2%, respectively. The ratio of AS events to AM events also showed an increasing trend under the RCP8.5 scenario. For example, the ratio would increase from 50.1% in 1981–2010 to 72% in the mid-period of 2070–2100 for the 24-month timescale, indicating 72% of the drought events in the mid-period of 2070–2100 are expected to be intense than the severe drought.

Evolutions of the duration and severity for all drought events under historical (1981–2010) and future RCP8.5 scenario periods (2010–2099) are shown in Figure 3-7. The drought duration and severity are predicted to be shorter and weaker in the mid-period of 2010–2040 than in the historical period, which is similar to past drought events. However, drought duration and severity show an increasing trend from mid-period of 2010–2040 to mid-period of 2070–2100 for all time scales. Compared with those in the mid-period of 2010–2040, the highest increasing rates for drought duration and drought severity in the mid-period of 2070–2100 are 62.2% (12-timescale) and 188.1% (24-timescale), respectively.

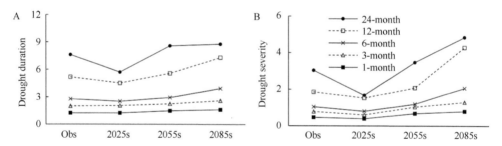

Figure 3-7 Temporal changes in average drought duration (A) and severity (B) for all drought events under RCP8.5 scenario. The "Obs" indicates the period of 1981–2010

As shown in Table 3-6, drought is projected to become more severe in the mid-period of 2070–2100. the intensity was found to be 43.3% (1-month), 61.5% (3-month), 62.5% (6-month), 86.3% (12-month), and 100% (24-month) higher in the mid-period of 2070–2100 than in mid-period of 2070–2100. Similarly, the average drought intensity during the mid-period of 2010–2040 was projected to be lower than that during the historical period. Based on these results, while drought components, such as the number of events, durations, and severity, tend to be lower in the mid-period of 2010–2040, drought risk is predicted to intensify in the mid-period of 2040–2070 and mid-period of 2070–2100, and will be severe in the mid-period of 2040–2070 and mid-period of 2070–2100 compared with the historical standards.

Table 3-6 Temporal changes in average drought intensity for all drought events in mid-period of 2010–2040, mid-period of 2040–2070, and mid-period of 2070–2100 under RCP8.5 scenario

Timescale	1981–2010	RCP8.5 scenario		
		mid-period of 2010–2040	mid-period of 2040–2070	mid-period of 2070–2100
1-month	0.37	0.32	0.42	0.46
3-month	0.34	0.26	0.39	0.42
6-month	0.31	0.24	0.33	0.39
12-month	0.26	0.22	0.27	0.41
24-month	0.25	0.17	0.24	0.34

3.4 Discussion

3.4.1 Trend variations between different drought indices

This work has shown that drought characteristics, including duration, intensity, and severity, have become moderate over the past 50 years in the Huang-Huai-Hai Plain based on the verified SPEI-PM index. These results are inconsistent with the previous studies, where Northern China was shown to have experienced a warm-drying trend (Wang et al., 2015a; Yu et al., 2014a). This inconsistency is likely due to the use of different indices and especially the variation arose from estimation of potential evapotranspiration with different indices. For example, PDSI-TH has been employed to detect global drying trends in the past decades (Dai, 2013; Nam et al., 2015). However, some studies used PM equation to calculate drought indices and concluded that drought has changed little worldwide (Sheffield et al., 2012) as well as in China over the past decades (Wang et al., 2015b). To estimate the variations between different drought indices in the Huang-Huai-Hai Plain, the Mann-Kendall trend test was performed on annual mean SPI, SPEI-TH, and SPEI-PM at 3-month scale under the historical (Figure 3-8) and future RCP8.5 (Figure 3-9) scenarios. Stations marked with red inverted triangles in both Figure 3-8 and Figure 3-9 are significantly drying areas, and stations marked with blue triangles are significantly wetting areas.

In general, the drought trend defined by SPI was driven by precipitation (McKee et al., 1993), while SPEI depends on both precipitation and ET_0 (Vicente-Serrano et al., 2010). For the historical period, precipitation decreased insignificantly by 1.01 mm·y^{-1} (Table 3-7). ET_0-TH increased by 1.32 mm·y^{-1}, which further aggravated water shortages. However, ET_0-PM decreased by 2.11 mm·y^{-1}, which made up for the increased precipitation. Thus, SPI and SPEI-TH showed a drying trend over the Huang-Huai-Hai Plain (Figure 3-8-A and Figure 3-8-B), while the SPEI-PM showed slightly wetter conditions. While precipitation is expected to increase by 1.88 mm·y^{-1} in the future, the

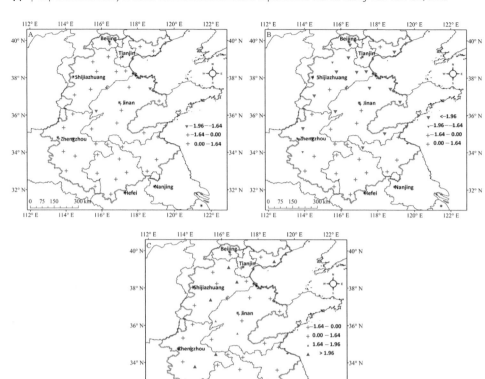

Figure 3-8 Trend variations of annual SPI-3 (A), SPEI-TH3 (B), and SPEI-PM3 (C) according to Mann-Kendall test and Theil-Sen's slope estimator in the Huang-Huai-Hai Plain during 1961–2010. These maps were obtained with the ArcGIS software

Table 3-7 Annual trend of precipitation and potential evapotranspiration (ET_0-PM and ET_0-TH) and four major climatic variables using the Mann-Kendall test and Theil-Sen's slope estimator

Item	Historical period (1961–2010)	RCP8.5 scenario (2010–2099)
Precipitation (mm·y^{-1})	−1.01	1.88*
ET_0-TH (mm·y^{-1})	1.32**	6.84**
ET_0-PM (mm·y^{-1})	−2.11**	3.58**
Daily temperature (°C·decade^{-1})	0.24**	0.77**
Relative humidity (%·y^{-1})	−0.05*	−0.05*
Wind speed (m·s^{-1}·y^{-1})	−0.01**	−0.00**
Net solar radiation (MJ·m^{-2}·y^{-1})	−12.12**	5.19**

*and ** represent the significant level at $P<0.05$ and $P<0.01$, respectively

amplification of ET_0 by both TH and PM equations counteracted increase (Table 3-7). Thus, SPI predicted wetter conditions in the future period (Figure 3-9-A). However,

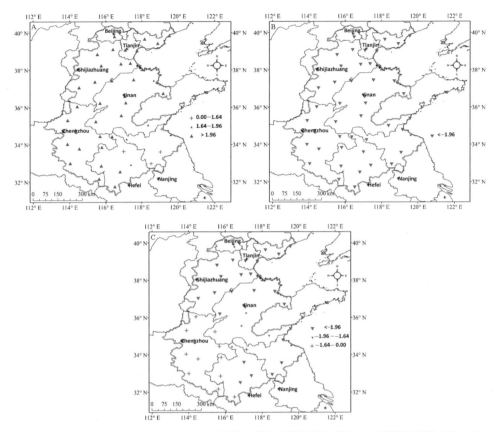

Figure 3-9 Trend variations of annual SPI-3 (A), SPEI-TH3 (B), and SPEI-PM3 (C) in the Huang-Huai-Hai Plain using Mann-Kendall test and Theil-Sen's slope estimator under future climate change (2010–2099). These maps were obtained with the ArcGIS software

SPEI-TH and SPEI-PM predicted that almost all meteorological stations would experience significant drying trends (Figure 3-9-B and Figure 3-9-C), except the southwest regions where SPEI-PM showed an insignificant trend. whoever conducts drought hazard assessment should understand these differences and the index chosen for drought studies should be justifiable.

3.4.2 Applicability of drought index

In this study, the applicability of SPI, SPEI-TH, and SPEI-PM was verified based on observed data from the agricultural drought areas in the Huang-Huai-Hai Plain. We found that SPEI-PM has higher correlation level with historical data. However, it does not mean that SPEI-PM would have the same applicability in other locations or systems. Remarkably, the best drought index for detecting impacts changes as a function of the analyzed system and the performance of the drought indices varied spatially (Vicente-Serrano et al., 2012). For example, the SPI is found to be well correlated with runoff anomaly in China (Zhai et al., 2010), while SPEI is better for hydrological application in western Canada.

For agricultural drought assessment, studies comparing the performance of several drought indices, like those evaluated here, would be preferable to determine the best drought index for identifying a certain drought type and its impacts on different systems. Nevertheless, the variable that could better reflect the drought impact for the analyzed system is of considerable importance. In this study, the observed data from drought area were used to evaluate the applicability of drought index for agricultural application. Several previous studies (Labudová et al., 2016; Potopova et al., 2015; Vicente-Serrano et al., 2012) have used detrended yield (or climatic yield), by applying the first-difference estimator or a linear regression model to eliminate the effects of technology change on the influence from actual crop yields, in order to assess the applicability of drought index. However, the reality of these studies was based on the hypothesis that yield fluctuations are mostly attributed to water stress, and agricultural technology has changed linearly (Yu et al., 2012).

Additionally, the comparison between SPEI-PM and SPEI-TH indicated that the way potential evapotranspiration is estimated would make differences in drought applicability and long-term drought trend. This difference has been found in other places of China (Wang et al., 2015b; Xu et al., 2015; Zhang et al., 2015b) and around the world (Begueria et al., 2014; Sheffield et al., 2012). The Thornthwaite (TH) and Penman-Monteith (PM) are widely used in drought index estimation. The TH model used for computing potential evapotranspiration in drought assessment is popularly used due to its simplicity and less data requirements (only temperature). Chen et al. (2005) concluded that the TH method overestimates ET_0 in southeast China where ET_0 is low, and underestimates in the northern and northwest parts where ET_0 is high when compared with pan data, and it does not follow the temporal variation well. Instead, PM equation is the most reliable estimation and is recommended by the FAO to calculate crop water requirements (Allen et al., 2005). Thus, considering the better applicability of SPEI-PM and estimation results, we recommend using the Penman-Monteith equation to calculate drought index in the Huang-Huai-Hai Plain.

3.5 Conclusions

The drought characteristics were investigated using "run theory" for both historical observation and future RCP8.5 scenarios in the Huang-Huai-Hai Plain based on the preferable drought index, i.e. SPEI-PM instead of SPI or SPEI-TH. The results can be summarized as follows:

(1) In the Huang-Huai-Hai Plain, FAO-56 Penman-Monteith (PM) formula based

SPEI, i.e. SPEI-PM, is more suitable for agricultural drought impact analysis as it has higher correlation coefficients with historical data than SPI and traditional SPEI.

(2) Based on calculations using the SPEI-PM, although total drought events showed no significant tendency over the historical period, the decreasing potential evapotranspiration reduced the drought duration, severity, and intensity from the 1960s to the 2000s.

(3) Compared with those during the historical period, drought characteristics, including the frequency, duration, severity, and intensity tend to be lower in the first thirty years in the future RCP8.5 scenario. However, drought condition is predicted to be intensified in the mid-period of 2040–2070 and the mid-period of 2070–2100, and will become very serious by historic standards after the mid-period of 2040–2070.

References

Ahlgren P, Bo J, Rousseau R. 2003. Requirements for a cocitation similarity measure, with special reference to Pearson's correlation coefficient. Journal of the Association for Information Science and Technology, 54(6): 550-560.

Allen C D, Macalady A K, Chenchouni H, Bachelet D, McDowell N, Vennetier M, Kitzberger T, Rigling A, Breshears D D, Hogg E H, Gonzalez P, Fensham R, Zhang Z, Castro J, Demidova N, Lim J H, Allard G, Running S W, Semerci A, Cobb N. 2010. A global overview of drought and heat-induced tree mortality reveals emerging climate change risks for forests. Forest Ecology and Management, 259(4): 660-684.

Allen R G, Pereira L S, Smith M, Raes D, Wright J L. 2005. FAO-56 dual crop coefficient method for estimating evaporation from soil and application extensions. Journal of Irrigation and Drainage Engineering-ASCE, 131(1): 2-13.

Begueria S, Vicente-Serrano S M, Reig F, Latorre B. 2014. Standardized precipitation evapotranspiration index (SPEI) revisited: Parameter fitting, evapotranspiration models, tools, datasets and drought monitoring. International Journal of Climatology, 34(10): 3001-3023.

Chen D L, Gao G, Xu C Y, Guo J, Ren G Y. 2005. Comparison of the Thornthwaite method and pan data with the standard Penman-Monteith estimates of reference evapotranspiration in China. Climate Research, 28: 123-132.

Chen H, Wang J, Huang J. 2014. Policy support, social capital, and farmers' adaptation to drought in China. Global Environmental Change, 24: 193-202.

Dai A. 2011a. Characteristics and trends in various forms of the Palmer Drought Severity Index during 1900–2008. Journal of Geophysical Research: Atmospheres, 116(D12): 115.

Dai A. 2011b. Drought under global warming: A review. Wiley Interdisciplinary Reviews: Climate Change, 2(1): 45-65.

Dai A. 2013. Increasing drought under global warming in observations and models. Nature Climate Change, 3: 52-58.

Gurrapu S, Chipanshi A, Sauchyn D, Howard A. 2014. Comparison of the SPI and SPEI on predicting drought conditions and streamflow in the Canadian prairies. Proceedings of the 28th Conference of Hydrology-94th American Meteorological Society Annual Meeting, 10: 2-6.

Hayes M, Svoboda M, Wall N, Widhalm M. 2011. The lincoln declaration on drought indices: universal meteorological drought index recommended. Bulletin of the American Meteorological Society, 92(4): 485-488.

Heim R R. 2002. A review of twentieth-century drought indices used in the United States. Bulletin of the American Meteorological Society, 83(8): 1149-1165.

Huang Y, Wang J, Jiang D, Zhou K, Ding X, Fu J. 2014. Surface water deficiency zoning of China based on surface water deficit index (SWDI). Water Resources, 41(4): 372-378.

Labudová L, Labuda M, Takáč J. 2016. Comparison of SPI and SPEI applicability for drought impact assessment on crop production in the Danubian Lowland and the East Slovakian Lowland. Theoretical and Applied Climatology, 128(1-2): 491-506.

Leng G, Tang Q, Rayburg S. 2015. Climate change impacts on meteorological, agricultural and hydrological droughts in China. Global and Planetary Change, 126: 23-34.

Li Y, Huang H, Ju H, Lin E, Xiong W, Han X, Wang H, Peng Z, Wang Y, Xu J, Cao Y, Hu W. 2015. Assessing vulnerability and adaptive capacity to potential drought for winter-wheat under the RCP8.5 scenario in the Huang-Huai-Hai Plain. Agriculture, Ecosystems & Environment, 209: 125-131.

Lu C, Fan L. 2013. Winter wheat yield potentials and yield gaps in the North China Plain. Field Crops Research, 143: 98-105.

McKee T B, Doesken N J, Kleist J. 1993. The relationship of drought frequency and duration to time scales. Proceedings of the 8th Conference on Applied Climatology, 17: 179-183.

Mei X, Kang S, Yu Q, Huang Y, Zhong X, Gong D, Huo Z, Liu E. 2013. Pathways to synchronously improving crop productivity and field water use efficiency in the North China Plain. Scientia Agricultura Sinica, 46(6): 1149-1157.

Ming B, Guo Y Q, Tao H B, Liu G Z, Li S K, Wang P. 2015. $SPEI_{PM}$-based research on drought impact on maize yield in North China Plain. Journal of Integrative Agriculture, 14(4): 660-669.

Mishra A K, Desai V R. 2005. Drought forecasting using stochastic models. Stochastic Environmental Research and Risk Assessment, 19(5): 326-339.

Nam W H, Hayes M J, Svoboda M D, Tadesse T, Wilhite D A. 2015. Drought hazard assessment in the context of climate change for South Korea. Agricultural Water Management, 160: 106-117.

Potopova V, Stepanek P, Mozny M, Tuerkott L, Soukup J. 2015. Performance of the standardised precipitation evapotranspiration index at various lags for agricultural drought risk assessment in the Czech Republic. Agricultural and Forest Meteorology, 202: 26-38.

Sentelhas P C, Gillespie T J, Santos E A. 2010. Evaluation of FAO Penman-Monteith and alternative methods for estimating reference evapotranspiration with missing data in Southern Ontario, Canada. Agricultural Water Management, 97(5): 635-644.

Sheffield J, Wood E F, Roderick M L. 2012. Little change in global drought over the past 60 years. Nature, 491(7424): 435-438.

Sillmann J, Kharin V V, Zwiers F W, Zhang X, Bronaugh D. 2013. Climate extremes indices in the CMIP5 multimodel ensemble: Part 2. Future climate projections. Journal of Geophysical Research-Atmospheres, 118(6): 2473-2493.

Sun J, Liu Y. 2013. Responses of tree-ring growth and crop yield to drought indices in the Shanxi province, North China. International Journal of Biometeorology, 58(7): 1521-1530.

Trenberth K E, Dai A, van der Schrier G, Jones P D, Barichivich J, Briffa K R, Sheffield J. 2014. Global warming and changes in drought. Nature Climate Change, 4(1): 17-22.

Vicente-Serrano S M, Begueria S, Lopez-Moreno J I. 2010. A Multiscalar drought index sensitive to global warming: The standardized precipitation evapotranspiration index. Journal of Climate, 23(7): 1696-1718.

Vicente-Serrano S M, Begueria S, Lopez-Moreno J I. 2011. Comment on "Characteristics and trends in various forms of the Palmer Drought Severity Index (PDSI) during 1900-2008" by Aiguo Dai. Journal of Geophysical Research-Atmospheres, 116(D19): 112.

Vicente-Serrano S M, Beguería S, Lorenzo-Lacruz J, Camarero J J, López-Moreno J I, Azorin-Molina C, Revuelto J, Morán-Tejeda E, Sanchez-Lorenzo A. 2012. Performance of drought indices for ecological, agricultural, and hydrological applications. Earth Interactions, 16(10): 1-27.

Wang H, Chen A, Wang Q, He B. 2015a. Drought dynamics and impacts on vegetation in China from 1982 to 2011. Ecological Engineering, 75: 303-307.

Wang L, Chen W. 2014. A CMIP5 multimodel projection of future temperature, precipitation, and climatological drought in China. International Journal of Climatology, 34(6): 2059-2078.

Wang W, Zhu Y, Xu R, Liu J. 2015b. Drought severity change in China during 1961–2012 Indicated by SPI and SPEI. Natural Hazards, 75(3): 2437-2451.

Wilhite D A, Sivakumar M V K, Pulwarty R. 2014. Managing drought risk in a changing climate: The role of national drought policy. Weather and Climate Extremes, 3: 4-13.

Wilhite D A, Svoboda M D, Hayes M J. 2007. Understanding the complex impacts of drought: A key to enhancing drought mitigation and preparedness. Water Resources Management, 21(5): 763-774.

Xu K, Yang D, Yang H, Li Z, Qin Y, Shen Y. 2015. Spatio-temporal variation of drought in China during 1961–2012: A climatic perspective. Journal of Hydrology, 526: 253-264.

Xu L, Wang H, Duan Q, Ma J. 2013. The temporal and spatial distribution of droughts during summer corn growth in Yunnan Province based on SPEI. Resources Science, 35(5): 1024-1034.

Yang J, Mei X, Huo Z, Yan C, Ju H, Zhao F, Liu Q. 2015. Water consumption in summer maize and winter wheat cropping system based on SEBAL model in Huang-Huai-Hai Plain, China. Journal of Integrative Agriculture, 14(10): 2065-2076.

Yevjevich V. 1967. An objective approach to definitions and investigations of continental hydrologic droughts. Hydrology Paper 23, Colorado State University, Fort Collins, CO: 18.

Yong B, Ren L, Hong Y, Gourley J J, Chen X, Dong J, Wang W, Shen Y, Hardy J. 2013. Spatial-temporal changes of water resources in a typical semiarid basin of North China over the past 50 years and assessment of possible natural and socioeconomic causes. Journal of Hydrometeorology, 14(4): 1009-1034.

Yu M, Li Q, Hayes M J, Svoboda M D, Heim R R. 2014a. Are droughts becoming more frequent or severe in China based on the Standardized Precipitation Evapotranspiration Index: 1951–2010? International Journal of Climatology, 34(3): 545-558.

Yu Q, Li L, Luo Q, Eamus D, Xu S, Chen C, Wang E, Liu J, Nielsen D C. 2014b. Year patterns of climate impact on wheat yields. International Journal of Climatology, 34(2): 518-528.

Yu Y, Huang Y, Zhang W. 2012. Changes in rice yields in China since 1980 associated with cultivar improvement, climate and crop management. Field Crops Research, 136(5): 65-75.

Zhai J, Su B, Krysanova V, Vetter T, Gao C, Jiang T. 2010. Spatial variation and trends in PDSI and SPI indices and their relation to streamflow in 10 large regions of China. Journal of Climate, 23(3): 649-663.

Zhang H L, Zhao X, Yin X G, Liu S L, Xue J F, Wang M, Pu C, Lal R, Chen F. 2015a. Challenges and adaptations of farming to climate change in the North China Plain. Climatic Change, 129(1-2): 213-224.

Zhang J, Sun F, Xu J, Chen Y, Sang Y, Liu C. 2015b. Dependence of trends in and sensitivity of drought over China (1961–2013) on potential evaporation model. Geophysical Research Letters, 43(1): 206-213.

Zhang X, Chen S, Sun H, Shao L, Wang Y. 2011. Changes in evapotranspiration over irrigated winter wheat and maize in North China Plain over three decades. Agricultural Water Management, 98(6): 1097-1104.

Zhang X Y, Chen S Y, Liu M Y, Pei D, Sun H Y. 2005. Improved water use efficiency associated with cultivars and agronomic management in the North China Plain. Agronomy Journal, 97(3): 783-790.

Chapter 4   Potential effect of drought on winter wheat yield using DSSAT-CERES-Wheat model over the Huang-Huai-Hai Plain, China

Abstract

The Huang-Huai-Hai Plain is recognized as a major crop-producing area that encompasses 19% of the total arable land in China and has experienced serious drought and water scarcity problems in recent years. In this chapter, the potential impacts of drought on wheat yield were determined at twelve stations representing different locations in the Huang-Huai-Hai Plain. The CERES-Wheat model was used to simulate the effects of a designed irrigation schedule on wheat yield. The cumulative probability of the yield reduction rate during the jointing to heading stage and the filling stage was investigated to determine the impacts of drought on the wheat yield during these two critical water demand periods. The results indicate that a significant gap between the actual and potential yields is visible, and can be attributed to the water stress and changes in the management inputs. The yield reduction rate for wheat was much higher during the jointing to heading stage than in the filling stage, with spatial variability due to the different meteorological conditions over the plain. The cumulative probability of the yield reduction rate was higher during the jointing to heading stage in the north region than in the south region. The findings from this study provide basic information for drought management and the rational irrigation of winter wheat in the Huang-Huai-Hai Plain by focusing on the potential effects of drought during the critical growth stages of winter wheat.

4.1   Introduction

Drought is one of the most damaging and widespread climate extremes, negatively affecting agricultural production, water resources, ecosystem function and human lives around the world (Dai, 2011; Dilley et al., 2005; Wilhite et al., 2007). In China, drought has had large impacts on agricultural production, seriously challenging the sustainability of the food supply and security. Additionally, with projected increasing temperatures and changing precipitation distributions, the drought characteristics are expected to change (Liu et al., 2017; Sillmann et al., 2013; Wang et al., 2015) and subsequently make crop production levels more uncertain. The Intergovernmental Panel on Climate Change (IPCC) has reported that the period of 1980–2012 was the hottest period in the past 1400 years, with the global temperature increasing by approximately 0.85°C in total (Gray, 2010). Although the enhanced atmospheric CO_2 level will increase the rate of photosynthesis, the adverse impacts resulting from the increasing temperature and changing water availability will probably outweigh the advantages of the higher CO_2 concentration (Wassmann et al., 2009), especially if the average temperature increases by more than 3°C (Attri and Rathore, 2003). Researchers have already reported a decline in the productivity of wheat caused by an increase in the temperature in China (Wilcox and Makowski, 2014; Yang et al., 2014).

The Huang-Huai-Hai Plain is a major crop producing area in China that encompasses 19% of the total arable land in China and has experienced serious drought and water scarcity problems in recent years (Yong et al., 2013), which has become the limiting factor for agricultural production (Zhang et al., 2015). Water limitations will likely be exacerbated in the future by increased food demand, soil quality deterioration and the over-exploitation of groundwater resources (Yang et al., 2015). Climate variability, especially extreme events such as drought, may result in a fluctuation of crop yields (Lu and Lan, 2013; Yu et al., 2011). Thus, understanding the potential impact of drought on crop yields is necessary to reduce the vulnerability of the current food production and establish mitigation strategies for agriculture in the Huang-Huai-Hai Plain.

Conventionally, earlier studies have focused on the yield reduction of winter wheat caused by water stress during the critical stage of water demand (Lv et al., 2007; Wu et al., 2002). Furthermore, the impact of drought on the crop yields was evaluated using a computer simulation experiment at a single station to support the irrigation practice. However, few research has evaluated the DSSAT model's performance in estimating the drought risk and drought impact on crop yields at typical growth stages. Drought is widely accepted to be the most serious agricultural meteorological disaster in the Huang-Huai-Hai Plain (Wang et al., 2016). Accordingly, there is an urgent demand for conducting a study to analyze and evaluate the effects of drought that lead to crop yields reductions in areas of limited water resources.

The DSSAT model has been widely used for yield gap analysis, decision making and planning, strategic and tactical management decisions and climate change studies in Asia (He et al., 2013; Jalota et al., 2014; Pathak et al., 2003; Singh and Thornton, 1992). The following objectives will be addressed in this research: (1) an investigation of the trends of irrigation requirements at 12 selected locations during the last 35 years; (2) the evaluation of the potential impact of drought on the winter wheat yield using DSSAT; and (3) the establishment of the probability distribution of the yield reduction at 12 typical sites over the Huang-Huai-Hai Plain. The results are expected to provide basic information for drought management and the rational irrigation of winter wheat in the Huang-Huai-Hai Plain by focusing on the potential effects of drought in the critical growth stages of winter wheat.

4.2 Materials and methods

4.2.1 Study region and data description

Study region

The Huang-Huai-Hai Plain is located in Northern China, extending over 31°14′–40°25′N and 112°33′–120°17′E, with 23.3 million ha of arable land (19% of the total in China), providing approximately 70% of the national wheat production and 30% of the national maize production with a dominant winter wheat-summer maize double-cropping system (Yang et al., 2015). The Huang-Huai-Hai Plain belongs to the extratropical monsoon climatic region. The annual mean precipitation is 500–800 mm, with more than 70% falling in July to September, and the atmospheric evaporative demand is approximately 1000 mm·y^{-1} (Zhang et al., 2011a). For the wheat-maize rotation system, rainfall can only meet 65% of the total agricultural water demand, and for winter wheat, only 25%–40% of the water demand is satisfied by rainfall (Mei et al., 2013). Irrigation water is primarily pumped from groundwater, and the groundwater level has decreased from a depth of 10 m in the 1970s to 32 m in 2001 and has continued to decrease at a rate of 1 m per year (Zhang et al., 2015). Thus, drought in this region not only challenges food supplies, but also results in a series of environmental problems.

Meteorological data

The daily weather data, including precipitation (mm), maximum and minimum temperatures (°C), wind speed (m·s^{-1}), relative humidity (%) and sunshine hours (h), recorded at meteorological observatories located over the region were collected from the China Meteorological Administration (Table 4-1). The district wheat yield data were collected from the Directorate of Wheat during the study period.

Table 4-1 Information of selected twelve stations in the Huang-Huai-Hai Plain

Station	Longitude	Latitude	Location	Precipitation (mm)	Temperature (°C)	Solar radiation (MJ·m^{-2})
Miyun	116°17′	39°56′	North	213.2	7.9	3684.8
Tianjin	117°43′	39°03′	North	192.6	9.5	3760.8
Luancheng	114°25′	38°02′	West	180.0	10.3	3748.7
Nangong	115°23′	37°22′	Central	178.2	10.5	3922.1
Yanzhou	116°51′	35°34′	Central	270.9	10.6	3893.9
Weifang	119°11′	36°45′	East	233.8	9.6	3997.1
Linyi	118°21′	35°03′	East	324.1	10.9	3993.6
Xinxiang	113°53′	35°19′	West	215.6	11.6	3769.3
Zhumadian	114°01′	33°	South	470.0	12.3	3606.3
Shangqiu	115°40′	34°27′	Central	315.1	11.5	3650.6
Xuzhou	117°29′	34°17′	Central	369.7	11.8	3859.7
Shou Xian	116°47′	32°33′	South	509.6	12.3	3778.3

Notes: The values described in this table were averages of precipitation, temperature, and solar radiation during the winter wheat growing season in 1981–2010

4.2.2 Calculation of precipitation deficit for winter wheat

In this study, the simulated irrigation amount was confirmed according to the practical situation and the precipitation deficit (PD) during the jointing to heading stage and the anthesis to milk stage.

$$PD = P_e - ET_c \qquad \text{Formula 4-1}$$

where, ET_c is the accumulative water requirement for winter wheat (mm) and P_e is the accumulative effective precipitation (mm).

Effective precipitation (P_e) is the amount of precipitation that is actually added and stored in the soil. During drier periods less than 5 mm of daily rainfall would not be considered effective, as this amount of precipitation would likely evaporate from the surface before soaking into the ground (Smith, 1992).

$$P_e = \begin{cases} P(4.17 - 0.2P)/4.17 & P < 8.3 \\ 4.17 + 0.1P & P \geqslant 8.3 \end{cases} \qquad \text{Formula 4-2}$$

where, P_e is the daily effective precipitation (mm·d^{-1}), P is the daily precipitation. The crop coefficients vary with the types of crops, as well as their growth stages and regions. In this study, the daily wheat coefficients in relation to the climatic conditions for the phonological phases of winter wheat were obtained following the approach developed by the FAO (Allen et al., 1998).

The irrigation schedule can be drawn through the investigation of the water consumption in different irrigation systems to make an appropriate irrigation programme under climate change (Cabelguenne et al., 1997; Santos et al., 2000). The simulated irrigation schedule was designed with four treatments in this study according to the practical situation on the basis of the same or other cultivation management measures. As shown in Table 4-2, CK is the treatment of full irrigation or no water stress and T1, T2, and T3 are the treatments with instead of no irrigation during the jointing and heading stage, filling stage and both stages, respectively.

Table 4-2 Simulated irrigation schedules based on fixed dates and precipitation deficit

Treatment	Irrigation volume in winter (mm)	Irrigation volume at jointing stage (mm)	Irrigation volume at filling stage (mm)
CK	60	PD1	PD2
T1	60	0	PD2
T2	60	PD1	0
T3	60	0	0

Notes: PD1 indicates precipitation deficit at the jointing to heading stage, PD2 indicates precipitation deficit at the filling stage

4.2.3 Crop model description

Crop models can be used to analyse the effects of various climatic factors on crop growth, DSSAT is an integrated software comprising different computer programs that can simulate crop growth and yield for research and decision-making. Its latest version DSSAT ver. 4.6.1.0 (http://dssat.net/downloads/dssat-v46) includes Cropping System Model (CSM), the primary modules of which are composed of weather module, soil module, soil-plant-atmosphere module, management operation module and 27 individual plant growth modules. Each plant growth model represents a different type of plant and simulates phenomenon, biomass and yield, based on the soil water and fertility supplement in response to weather and management (Jones et al., 2003). The model offers options to change the irrigation schedule to simulate possible climatic drought effects on the crop. The DSSAT Crop Environment Resource Synthesis (CERES)-wheat model (He et al., 2013; 2012) was adopted here to simulate the wheat yield during 1981–2014 at all twelve locations.

4.2.4 Statistical tests for trend analysis

Recent studies suggest that non-parametric tests are more suitable for non-normally distributed and censored data, which are frequently encountered in meteorological and hydrological time series. Among them, the non-parametric Mann-Kendall test (Kendall, 1975; Mann, 1945) is a widely accepted method for trend analysis in hydrological and meteorological series (Jhajharia et al., 2012; Zheng and Wang, 2014), with little effect from the presence of outliers in the data. In addition, it is highly recommended for general use by the World Meteorological Organization (Zhang et al., 2011b). Therefore, we used the Mann-Kendall test for the trend analysis of our data. The magnitudes of the trends in the time series were estimated using the non-parametric Theil-Sen's slope (Sen, 1968; Theil, 1992), which is robust because it is resistant to the effects of outlier values in the observations (Jhajharia et al., 2015; Su et al., 2015).

4.3 Results

4.3.1 DSSAT evaluation

The DSSAT model (CERES-Wheat) was calibrated and validated by testing different genetic coefficients (Table 4-3) to simulate the observed phenological growth and yield of wheat for approximately five years over the Huang-Huai-Hai Plain. The simulation of the DSSAT model shows advances or delays of crop duration, anthesis and maturity periods for less than 5 days compared with the observed data. Figure 4-1 presents a comparison of the observed and simulated wheat yields for the entire calibration and validation period. For the simulation of wheat yield, the coefficient of determination (R^2) is 0.81, and the normalized root-mean-square error (NRMSE) is estimated to be, on average, 8.5 kg·ha^{-1} for wheat. The overall results show an appropriate capability of the model to

Table 4-3 Statistics of observed and simulated dates of flowering and maturity stage and yields

Station	Cultivar	P1V (d)	P1D (%)	P5 (°C·d)	G1 (no·g^{-1})	G2 (mg)	G3 (g)	PHINT (°C·d)	NRMSE Anthesis (d)	NRMSE Maturity (d)	NRMSE Yield (kg·ha^{-1})
Miyun	Jingdong8	58.34	31.65	536.30	25.75	24.80	1.18	95	0.27	0.52	9.64
Tianjin	Jing9428	44.23	72.63	513.20	23.86	20.46	1.91	95	2.47	0.51	3.08
Shijiazhuang	Shixin733	11.38	75.43	708.30	18.80	35.29	1.02	95	0.61	0.74	7.79
Nangong	Han6172	22.84	53.22	462.60	23.51	26.55	1.05	95	0.49	1.02	10.00
Yanzhou	Jining16	21.87	50.31	579.40	15.80	45.54	1.09	95	0.42	0.90	11.56
Weifang	Jimai22	6.42	61.77	559.30	27.19	56.80	1.01	95	0.67	1.09	9.08
Linyi	Linmai4	44.36	55.41	593.5	25.89	30.68	1.307	95	0.36	0.68	7.39
Xinxiang	Xinmai6	41.67	53.09	554.40	25.68	23.90	1.15	95	0.54	0.77	9.66
Zhumadian	Zhengmai9023	39.78	4.59	604.50	16.67	62.84	1.00	95	0.68	1.14	3.08
Shangqiu	Wenmai8	61.93	18.46	548.60	26.21	33.92	1.12	95	0.74	0.26	10.03
Xuzhou	Xuzhou25	36.44	59.54	614.90	24.23	28.86	1.01	95	0.88	0.55	11.57
Shou Xian	Wanmai27	55.55	15.83	524.10	15.23	46.41	1.01	95	1.46	0.97	9.00

Notes: P1V is the vernalization parameter (d). P1D is the photoperiod parameter (%). P5 is grain filling duration parameter (°C·d). G1 is the grain parameter at anthesis (number·g^{-1}). G2 is the grain filling rate parameter (mg). G3 is the dry weight of a single stem and spike (g). PHINT is interval between successive leaf tip appearances (°C·d)

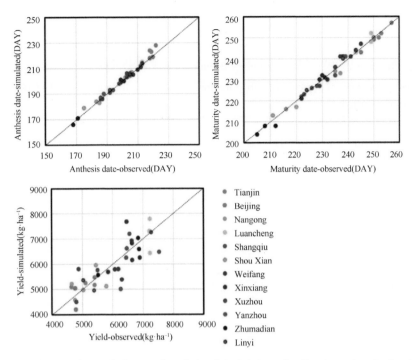

Figure 4-1 Comparison of observed and simulated dates of anthesis and maturity stage (d) and yields (kg·ha^{-1}) for the entire calibration and validation period

simulate the yield of winter wheat, and therefore it may be used to simulate the wheat yield to analyse the various aspects of climatic drought (Wen and Chen, 2011).

4.3.2 Trends and persistence of typical growth date and precipitation deficit

The annual magnitudes of the jointing date, heading date, anthesis date and milk ripe date for winter wheat at the selected twelve stations from 1981 to 2014 are described in Table 4-4. The linear regression is characterized by the fact that the jointing date had a declining trend for all the selected stations. The jointing date occurred earlier: −2.4 d·decade^{-1} and −4.0 d·decade^{-1} by the M-K test for the north and west stations, respectively, in the past 34 years. Similarly, for the central, east and south stations, significant shifts to earlier jointing of −3.4 d·decade^{-1}, −1.1 d·decade^{-1} and −7.4 d·decade^{-1}, respectively, were detected. Consequently, the heading date also shifted by −3.0 d·decade^{-1} for both the north and west stations, while the anthesis date shifted by −2.9 d·decade^{-1} and −1.7 d·decade^{-1}, respectively. Furthermore, for the central, east and south stations, significant decreases in magnitude of −3.3 d·decade^{-1}, −1.3 d·decade^{-1} and −7.1 d·decade^{-1}, respectively, were found, and the values were 2.7 d·decade^{-1}, −0.3 d·decade^{-1} and −5.9 d·decade^{-1} for anthesis. Finally, the milk ripe date also shifted by

−1.9 d·decade^{-1} and −1.8 d·decade^{-1} in the north and south stations, while it shifted to later dates in the west, central and east stations. The largest decrements were found for the jointing date, followed by the heading date and anthesis date.

Table 4-4 Variation of typical growth date for wheat and precipitation deficit (PD) during typical growth stages in 1981–2010

Station	Jointing date		Heading date		Anthesis date		Milk ripe date		PD during jointing to heading stage	PD during anthesis to milk ripe stage
	Mean (d)	Slope (d·decade^{-1})	Mean (d)	Slope (d·decade^{-1})	Mean (d)	Slope (d·decade^{-1})	Mean (d)	Slope (d·decade^{-1})	Mean (mm)	Mean (mm)
Miyun	113	−4.0**	132	−2.4**	137	−2.6**	154	−1.5**	−73.9	−48.1
Tianjin	108	−0.8**	128	−3.6**	133	−3.1**	148	−2.2**	−77.3	−26.7
Shijiazhuang	99	−2.4**	120	−2.6**	127	−1.8**	140	−1.2**	−78.2	−38.5
Nangong	94	0.5**	117	−2.9**	124	−2.2**	141	−0.3**	−91.0	−53.6
Yanzhou	90	−4.2**	114	−3.1**	121	−3.1**	142	1.4	−72.8	−43.6
Weifang	101	−0.7	122	−0.7**	128	1.2**	147	1.2**	−85.9	−36
Linyi	97	−1.4	120	−1.9**	127	−1.7**	147	0.7**	−77.3	−45.3
Xinxiang	87	−5.5**	113	−3.3**	120	−1.6**	138	1.3**	−53.7	−55.2
Zhumadian	79	−7.9**	105	−6.6**	112	−6.0**	134	−0.8**	−27.2	−41.4
Shangqiu	86	−3.5**	111	−3.6**	117	−2.5**	140	1.8*	−59	−39.6
Xuzhou	82	−6.4**	113	−3.4**	120	−3.1**	139	−1.1**	−72.1	−36
Shou Xian	83	−6.8**	109	−7.5**	116	−5.8**	134	−2.7**	−29.3	−41.6

Notes: The slope (d·decade^{-1}) means the temporal trend in jointing, heading, anthesis and milk ripe date of winter wheat in 1981–2010 using Mann-Kendall test and Theil-Sen's slope estimator

As indicated in Table 4-4, the precipitation deficit (−66.5 mm on average) during the jointing to heading stage was slightly lower than that (−42.1) during the anthesis to milk ripe stage. Smaller fluctuations were found during the anthesis to milk ripe stage at the 12 selected stations, with a range of 20–40 mm. The precipitation deficit reached its lowest values in the north, east and central stations during the jointing to heading stage and in the west and central stations during the filling stage. The largest values for both were obtained from two southern stations. The value of the precipitation deficit is the input of the DSSAT simulation. Santos et al. (2000) reported that the investigation of the water demand under different irrigation systems could serve the purpose of determining the optimal irrigation schedule.

4.3.3 Variation of yield reduction rate

Using DSSAT, we determined the annual variation of potential yield reduction for wheat caused by drought stress during the jointing to heading stage and filling stage at the selected

12 stations in a recent 30-year period across the Huang-Huai-Hai Plain (Figure 4-2). Strong fluctuations were found in the yield reduction in Tianjin, with two periods of low values in the late 1980s and 1990s. For the Linyi station, the yield reduction rate declined in 1980–2000, but increased in 2001–2010. It can be seen from Figure 4-3 that the largest yield reduction was found in the middle of the 1990s for Tianjin and Shijiazhuang stations and in the late 1990s for other four stations, mainly due to drought events in 2000s over the Huang-Huai-Hai Plain. In the 1980s, the yield reduction rate declined significantly for all selected stations during the jointing to heading stage. During the filling stage, the rate declined except in Tianjin and shijiazhuang. Furthermore, the yield reduction rate increased for the Tianjin and Linyi stations during the jointing to heading stage, while it decreased for Shangqiu station in the 1990s. For the filling stage, the yield reduction rate declined in only the Tianjin station, and it increased in the Zhumadian and Shangqiu stations. To summarize, the trends in the yield reduction rate without irrigation in both the jointing to heading stage and filling stage were similar to those of the yield reduction due to drought in the jointing to heading stage.

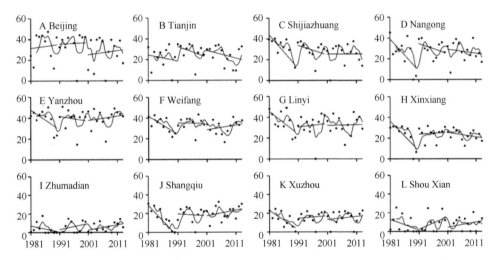

Figure 4-2 Variation of simulated yield reduction (%) for wheat under the treatment of no irrigation during the jointing and heading stage (T1 treatment)

4.3.4 Cumulative probability of yield reduction rate

As described in Figure 4-5, the cumulative probabilities of the yield reduction rate during the jointing to heading stage and the filling stage were investigated to determine the impacts of drought on the wheat yield in these two critical water demand periods. It can be seen that the yield reduction rate under the same level was much higher during

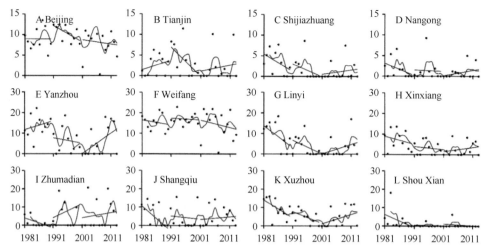

Figure 4-3 Variation of simulated yield reduction (%) for wheat under the treatment of no irrigation during the filling stage (T2 treatment)

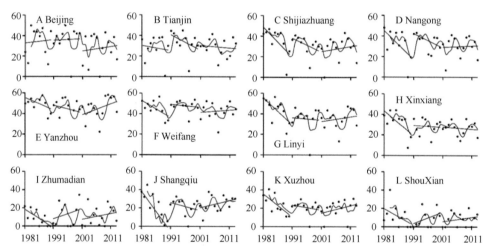

Figure 4-4 Variation of simulated yield reduction (%) for wheat under the treatment of no irrigation during both jointing-heading stage and filling stage (T3 treatment)

the jointing to heading stage than in the filling stage. The cumulative probability of the yield reduction rate, which is over 60% during the jointing to heading stage, was approximately 15% in the north and west stations and turned out to be around 65% in the north and east stations when the yield reduction rate was over 40%. The cumulative probability of the yield reduction rate was higher during the jointing to heading stage in the north part than in the south region, and almost the same during the filling stage.

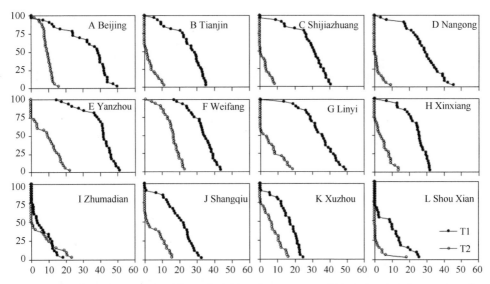

Figure 4-5 Cumulative probability of yield reduction rate (%) during jointing to heading stage (T1) and filling stage (T2) for winter wheat, respectively, in 12 typical sites

4.4 Discussion

Evapotranspiration is widely acknowledged as a factor that determines climatic drought in arid and semi-arid regions, where minor variations in precipitation or temperature easily induce significant changes in hydrological processes and crop water requirements (Ma et al., 2004; Temesgen et al., 2005). Over the last decades, the aridity index is expected to have different trends in various regions (Liu et al., 2013a; Su et al., 2015; Wu et al., 2006). Droughts have become more frequent and intense during the last decades, which presenting a direct threat to crop growth in vast areas across China (Dalin et al., 2015; Wang et al., 2016; 2011). Liu et al. (2013b) identified an increasing trend of total water deficit of winter wheat in the past three decades in Northern China; therefore, a negative influence on wheat production and water resources in the area can be expected in the future if no management changes, such as irrigation techniques, cropping pattern, mulching, etc., are adopted (Allen et al., 1998; Olesen and Bindi, 2002). The present study gives a first hint of yield fluctuations based on the CERES-Wheat model to be expected and regions to should be prioritized. The yield reduction rate for wheat was much higher in the jointing to heading stage than in the filling stage, which is in agreement with experimental results in Anhui province (Wang et al., 2001) and simulated findings by WOFOST model in Zhengzhou agro-meteorological station (Zhang et al., 2012). The drought reduced winter wheat yield by affecting grain-filling intensity due to soil water deficit. The reduction in yield was significantly different when drought occurred at different developmental stages.

There was a higher impact on winter wheat when drought occurred at several developmental stages than at a single developmental stage. A study in North China revealed that under wheat production, mean daily soil evaporation from less and additional mulching land was reduced by 16% and 37%, respectively (Chen et al., 2007). In Spain, Döll (2002) has predicted a declining in irrigation requirements by 2020 on account of the possibility of earlier sowing under more favourable higher temperatures. It will be necessary to develop feasible straw (film) mulching, regulated deficit irrigation, and soil water storage and preservation to reduce pressure on groundwater over-exploitation, especially for winter wheat in the Huang-Huai-Hai Plain.

Process-based crop models have been widely used to assess the impacts of future climate change on crop production, but its applications on impact estimation of past climate change are limited because the model's capacity to reproduce the actual yield responses to environment largely depends on a few important procedures such as calibration, validation and data aggregation. Furthermore, the simulation results often suffer from a number of uncertainties from pests, disease and some largely underestimated extreme events. Nonetheless, our results are still in line with some previous studies (Lv et al., 2007; Zhang et al., 2012). The reduction of the grain number is due to water stress during the jointing to heading stage, while the reduction of the grain weight is mainly caused by water stress during the filling stage (Huang et al., 2013; Sun et al., 2004).

4.5 Conclusions

We have examined the impact of drought on the wheat yield at twelve different locations in the Huang-Huai-Hai Plain. The study provides insight into the impact of irrigation scheduling on the yield of winter wheat, the most important cereal crop in the Huang-Huai-Hai Plain. Through DSSAT simulations, we showed that the wheat yields vary spatially due to the same treatment all over the Huang-Huai-Hai Plain. A gap between the actual and potential yield is significant which can be attributed to the water stress and changes in the management inputs. The yield reduction rate for wheat was much high in the jointing to heading stage than in the filling stage, with spatial variability due to the climate conditions. The cumulative probability of the yield reduction rate during the jointing to heading stage was higher in the north part than in the southern region.

References

Allen R G, Pereira L S, Raes D, Smith M. 1998. Crop evapotranspiration-Guidelines for computing crop water requirements. Irrigation and Drainage Paper 56 Food and Agriculture Organization of the United Nations, Rome.

Attri S D, Rathore L S. 2003. Simulation of impact of projected climate change on wheat in India. International Journal of Climatology, 23(6): 693-705.

Cabelguenne M, Debaeke P, Puech J, Bosc N. 1997. Real time irrigation management using the EPIC-PHASE model and weather forecasts. Agricultural Water Management, 32(3): 227-238.

Chen S Y, Zhang X Y, Pei D, Sun H Y, Chen S L. 2007. Effects of straw mulching on soil temperature, evaporation and yield of winter wheat: field experiments on the North China Plain. Annals of Applied Biology, 150(3): 261-268.

Dai A. 2011. Drought under global warming: a review. Wiley Interdisciplinary Reviews Climate Change, 2(1): 45-65.

Dalin C, Qiu H, Hanasaki N, Mauzerall D L, Rodriguezitrube I. 2015. Balancing water resource conservation and food security in China. Proceedings of the National Academy of Sciences of the United States of America, 112(15): 4588-4593.

Dilley M, Chen R S, Deichmann U, Lerner Lam A L, Arnold M. 2005. Natural disaster hotspots: a global risk analysis. Vol. 5, World Bank Publications, Washington, DC.

Döll P. 2002. Impact of climate change and variability on irrigation requirements: A global perspective. Climatic Change, 54(3): 269-293.

Gray V. 2010. Climate Change 2013–The Physical Science Basis. South African Geographical Journal Being a Record of the Proceedings of the South African Geographical Society, 92: 86-87.

He J, Cai H, Bai J. 2013. Irrigation scheduling based on CERES-Wheat model for spring wheat production in the Minqin Oasis in Northwest China. Agricultural Water Management, 128: 19-31.

He J, Dukes M D, Hochmuth G J, Jones J W, Graham W D. 2012. Identifying irrigation and nitrogen best management practices for sweet corn production on sandy soils using CERES-Maize model. Agricultural Water Management, 109: 61-70.

Huang L, Gao Y, Qiu X, Li X. 2013. Effects of irrigation amount and stage on yield and water consumption of different winter wheat cultivars. Transactions of the CSAE, 29(14): 99-108.

Jalota S K, Vashisht B B, Kaur H, Kaur S, Kaur P. 2014. Location specific climate change scenario and its impact on rice and wheat in Central Indian Punjab. Agricultural Systems, 131: 77-86.

Jhajharia D, Dinpashoh Y, Kahya E, Singh V P, Fakheri-Fard A. 2012. Trends in reference evapotranspiration in the humid region of northeast India. Hydrological Processes, 26(3): 421-435.

Jhajharia D, Kumar R, Dabral P P, Singh V P, Choudhary R R, Dinpashoh Y. 2015. Reference evapotranspiration under changing climate over the Thar Desertin India. Meteorological Applications, 22(3): 425-435.

Jiang Y, Zhang L, Zhang B, He C, Jin X, Bai X. 2016. Modeling irrigation management for water conservation by DSSAT-maize model in arid northwestern China. Agricultural Water Management, 177: 37-45.

Jones J W, Hoogenboom G, Porter C H, Boote K J, Batchelor W D, Hunt L A, Wilkens P W, Singh U, Gijsman A J, Ritchie J T. 2003. DSSAT cropping system model. European Journal of Agronomy, 18(3-4): 235-265.

Ju H, Xiong W, Xu Y L, Lin E D. 2005. Impacts of Climate Change on Wheat Yield in China. Acta Agronomica Sinica. 31(10): 1340-1343.

Kendall M. 1975. Rank Correlation Methods. London: Charles Griffin & Company Ltd.

Liu Q, Mei X R, Yan C, Ju H, Yang J. 2013b. Dynamic variation of water deficit of winter wheat and its possible climatic factors in Northern China. Acta Ecologica Sinica, 33(20): 6643-6651.

Liu Q, Yan C, Ju H, Garre S. 2017. Impact of climate change on potential evapotranspiration under a historical and future climate scenario in the Huang-Huai-Hai Plain, China. Theoretical and Applied Climatology, 132(1-2): 387-401.

Liu X, Zhang D, Luo Y, Liu C. 2013a. Spatial and temporal changes in aridity index in northwest China: 1960 to 2010. Theoretical and Applied Climatology, 112(1-2): 307-316.

Lu C, Lan F. 2013. Winter wheat yield potentials and yield gaps in the North China Plain. Field Crops Research, 143: 98-105.

Lv L, Hu Y, Li Y, Wang P. 2007. Effect of irrigating treatments on water use efficiency and yield of different wheat cultivars. Journal of Triticeae Crops, 27: 88-92.

Ma Z, Dan L, Hu Y. 2004. The extreme dry/wet events in Northern China during recent 100 years. Journal of Geographical Sciences, 14(3): 275-281.

Mann H B. 1945. Nonparametric test against trend. Econometrica, 13(3): 245-259.

Mei X R, Kang S Z, Qiang Y U, Huang Y F, Zhong X L, Gong D Z, Huo Z L, Liu E K. 2013. Pathways to synchronously improving crop productivity and field water use efficiency in the North China Plain. Scientia Agricultura Sinica, 46(6): 1149-1157.

Olesen J E, Bindi M. 2002. Consequences of climate change for European agricultural productivity, land use and policy. European Journal of Agronomy, 16(4): 239-262.

Pathak H, Ladha J K, Aggarwal P K, Peng S, Das S, Singh Y, Singh B, Kamra S K, Mishra B, Sastri A S R A S. 2003. Trends of climatic potential and on-farm yields of rice and wheat in the Indo-Gangetic Plains. Field Crops Research, 80(3): 223-234.

Santos A M, Cabelguenne M, Santos F L, Oliveira M R, Serralheiro R P, Bica M A. 2000. EPIC-PHASE: a model to explore irrigation strategies. Journal of Agricultural Engineering Research, 75(4): 409-416.

Sen P K. 1968. Estimates of the regression coefficient based on Kendall's tau. Journal of the American Statistical Association, 63(324): 1379-1389.

Sillmann J, Kharin V V, Zhang X, Zwiers F W, Bronaugh D. 2013. Climate extremes indices in the CMIP5 multimodel ensemble: Part 1. Model evaluation in the present climate. Journal of Geophysical Research Atmospheres, 118(4): 1716-1733.

Singh U, Thornton P K. 1992. Using crop models for sustainability and environmental quality assessment. Outlook on Agriculture, 21(3): 209-218.

Smith M. 1992. CROPWAT: A computer program for irrigation planning and management-irrigation and drainage paper 46. Rome: Food and Agriculture Organization of the United: 20-21.

Su X, Singh V P, Niu J, Hao L. 2015. Spatiotemporal trends of aridity index in Shiyang River basin of northwest China. Stochastic Environmental Research and Risk Assessment, 29(6): 1571-1582.

Sun B, Wang Y, Li X, Liu F. 2004. The influence of climate and cultivation on the yield components of winter wheat in different years. Journal of Triticeae Crops, 24(2): 83-87.

Temesgen B, Eching S, Davidoff B, Frame K. 2005. Comparison of some reference evapotranspiration equations for California. Journal of Irrigation & Drainage Engineering, 131(1): 73-84.

Theil H. 1992. A rank-invariant method of linear and polynomial regression analysis. Nederl. Akad. Wetensch. Proc, 12(2): 345-381.

Wang A, Lettenmaier D P, Sheffield J. 2011. Soil moisture drought in China, 1950–2006. Journal of

Climate, 24(13): 3257-3271.

Wang H, Chen A, Wang Q, He B. 2015. Drought dynamics and impacts on vegetation in China from 1982 to 2011. Ecological Engineering, 75: 303-307.

Wang H, Vicente-Serrano S M, Tao F, Zhang X, Wang P, Zhang C, Chen Y, Zhu D, Kenawy A E. 2016. Monitoring winter wheat drought threat in Northern China using multiple climate-based drought indices and soil moisture during 2000–2013. Agricultural & Forest Meteorology, s 228-229: 1-12.

Wang M, Zhang C, Yao W, Wang X. 2001. Effects of drought stress in different development stages on wheat yield. Journal of Anhui Agricultural Sciences, 29(5): 605-607, 610.

Wassmann R, Jagadish S V K, Sumfleth K, Pathak H, Howell G, Ismail A, Serraj R, Redona E, Singh R K, Heuer S. 2009. Regional Vulnerability of Climate Change Impacts on Asian Rice Production and Scope for Adaptation. Advances in Agronomy, 102(9): 91-133.

Wen X, Chen F. 2011. Simulation of climatic change impacts on yield potential of typical wheat varieties based on DSSAT model. Transactions of the CSAE, 27(14): 74-79.

Wilcox J, Makowski D. 2014. A meta-analysis of the predicted effects of climate change on wheat yields using simulation studies. Field Crops Research, 156: 180-190.

Wilhite D A, Svoboda M D, Hayes M J. 2007. Understanding the complex impacts of drought: A key to enhancing drought mitigation and preparedness. Water Resources Management, 21(5): 763-774.

Wu S, Gao H, Wang S, Duan G. 2002. Analysis on the effect of drought on the grain weight grow and the character of the grain filling of winter wheat. Agricultural Research in the Arid Areas, 20(2): 49-51.

Wu S, Yin Y, Zheng D, Yang Q. 2006. Moisture conditions and climate trends in China during the period 1971–2000. International Journal of Climatology, 26(2): 193-206.

Yang J M, Yang J Y, Dou S, Yang X M, Hoogenboom G. 2013. Simulating the effect of long-term fertilization on maize yield and soil C/N dynamics in northeastern China using DSSAT and CENTURY-based soil model. Nutrient Cycling in Agroecosystems, 95(3): 287-303.

Yang J, Mei X, Huo Z, Yan C, Zhao F, Liu Q. 2015. Water consumption in summer maize and winter wheat cropping system based on SEBAL model in Huang-Huai-Hai Plain, China. Journal of Integrative Agriculture, 14(10): 2065-2076.

Yang P, Wu W, Li Z, Yu Q, Inatsu M, Liu Z, Tang P, Zha Y, Kimoto M, Tang H. 2014. Simulated impact of elevated CO_2, temperature, and precipitation on the winter wheat yield in the North China Plain. Regional Environmental Change, 14(1): 61-74.

Yang Y, Yang Y, Moiwo J P, Hu Y. 2010. Estimation of irrigation requirement for sustainable water resources reallocation in North China. Agricultural Water Management, 97(11): 1711-1721.

Yong B, Ren L, Hong Y, Gourley J J, Chen X, Dong J, Wang W, Shen Y, Hardy J. 2013. Spatial-temporal changes of water resources in a typical semiarid basin of North China over the past 50 years and assessment of possible natural and socioeconomic causes. Journal of Hydrometeorology, 14: 1009-1034.

Yu M, Hayes M J, Heim R R, Li Q. 2011. Are droughts becoming more frequent or severe in China based on the Standarized Precipitation Evapotran-spiration Index: 1951-2010? International Journal of Climatology, 34(3): 545-558.

Zhang H L, Zhao X, Yin X G, Liu S L, Xue J F, Wang M, Pu C, Lal R, Chen F. 2015. Challenges and adaptations of farming to climate change in the North China Plain. Climatic Change, 129(1-2): 213-224.

Zhang J, Zhao Y, Wang C, Yang X. 2012. Impact simulation of drought disaster at different developmental stages on winter wheat grain-filling and yield. Chinese Journal of Eco-Agriculture, 20(9): 1158-1165.

Zhang Q, Xu C-Y, Chen X. 2011b. Reference evapotranspiration changes in China: natural processes or human influences? Theoretical and Applied Climatology, 103(3-4): 479-488.

Zhang X, Chen S, Sun H, Shao L, Wang Y. 2011a. Changes in evapotranspiration over irrigated winter wheat and maize in North China Plain over three decades. Agricultural Water Management, 98(6): 1097-1104.

Zheng C, Wang Q. 2014. Spatiotemporal variations of reference evapotranspiration in recent five decades in the arid land of Northwestern China. Hydrological Processes, 28(25): 6124-6134.

Chapter 5   Investigation of the impact of climate change on wheat yield using DSSAT-CERES-Wheat model over the Huang-Huai-Hai Plain, China

Abstract

Progress in understanding the impact of climate change on crop yields is essential for agricultural climate adaptation, especially for the Huang-Huai-Hai Plain, which is the grain production base of China. In this study, the effects of climate change (including solar radiation, maximum temperature, minimum temperature, and precipitation) between historical records (1985–2014) and future climate projections under the RCP4.5 and RCP8.5 pathways (2021–2050) on winter wheat yield were analyzed using the CERES-Wheat crop model. The results show that solar radiation (SRAD), maximum temperature (TMAX), minimum temperature (TMIN), and precipitation (PRCP) increased during the winter wheat growing seasonal. The average SRAD received during the wheat growing seasons between 2021–2050 increased by 4.60% and 3.82% under the RCP4.5 and RCP8.5 scenarios, respectively, as compared to 1985–2014. The TMAX and TMIN of the wheat growing seasons during 2021–2050 were 1.30°C and 1.07°C higher under RCP4.5 and 1.28°C and 1.09°C higher under RCP8.5. Precipitation during this season also increased by 15.01% and 16.44% under the two pathways. According to our model simulations, these changes will have an overall positive impact on yield which would increase by 17.34% and 15.76% under RCP4.5 and RCP8.5, respectively. We isolated the effect of each single weather variable on crop yields while keeping the other variables at the level of the historical data, and found that the results differed from one another. Increases in temperature and precipitation result in positive evolutions: increases of yield by 5.45% and 14.79%, respectively, under RCP4.5, and by 6.15% and 14.16%, respectively, under

RCP8.5. Interestingly, the supposed positive impact of increasing solar radiation is visible in our simulations. Higher solar radiation also increases the evaporation amount and water stress and consequently reduces the photosynthetically active radiation conversion to dry matter ratio (PARUE). Logically, these negative impacts are lower at southern stations of the Huang-Huai-Hai Plain with less water stress. Thus, in conclusion, to take full advantage of increasing SRAD, TMAX, TMIN and PRCP, the optimized management of agricultural water resources and the improvement of water utilization efficiency (WUE) are essential, especially for the Huang-Huai-Hai Plain, which has already suffered from serious water shortages.

5.1 Introduction

Agro-ecosystems are very sensitive to the effects of climate change, as they are directly exposed to changes in the atmosphere (Piao et al., 2010). Global warming increases the availability of the thermal resources in high-latitude areas so that the area of arable land has expanded (Ju et al., 2013); however, it worsens the severity of heat and drought stress for arid and semi-arid regions, which causes the food supply to fluctuate (Yuan et al., 2016). China is a large, densely populated country, so assuring food security is one of the priorities of the government. Thus, investigation of the impacts of climate change, and how to benefit from or mitigate its effects, is of great importance for agricultural adaptation and national food security (Ju et al., 2013; Piao et al., 2010).

Statistical models and numerical simulations are two popular approaches that have been widely used to investigate climate-crop relationships (Shi et al., 2013). Based on recorded yields and climate data, time-series regression models at the station single level can be established, and the signs of the regression coefficients represent the negative or positive impacts of climate change (David and Christopher, 2007; Tao et al., 2008). In recent years, statistical models have been witnessed to be adopted in investigating crop responses to climatic warming more than time series-based statistical models (Schlenker and Roberts, 2009; Tao et al., 2014; Wolfram and David, 2010). However, for these statistical models, the influences of non-climatic factors (such as cultivar and fertilizer change) on yield fluctuations need to be eliminated before establishing the statistical model. Although some simple elimination methods, such as the first-difference yield and the linear detrended yield, have been widely used in several studies, they are all based on certain hypotheses. For example, the linear detrended method is based on the concept that crop management and variety renewal change linearly with time, which does not match the practices of actual crop growth in the field (Xiong et al., 2014). Thus, the validity of these results remains questionable, and even conflicting results have been obtained using different elimination methods (Shi et al., 2013). In recognizing the complex interactions between crop growth and environmental factors, numerical simulations, or crop models, have become popular research tools for researchers in agro-meteorology in recent years. Dynamic crop models, such as the Erosion Productivity Impact Calculator (EPIC) model (Williams et al., 1983), APSIM (Keating et al., 2003), and DSSAT-CERES (Jones et al., 1998), have been tested and used in quantifying responses to water, nitrogen and weather at scales ranging from fields to regions around the world. Of these models, DSSAT is a software platform that includes multiple models for different crops, and it can quantitatively predict the growth and production of annual field crops given the interactions among the atmospheric and soil environments, cultivar factors and management (Ritchie et al., 1998). DSSAT has been used for climate change and climate extreme impact assessment for rice, wheat and maize in different zones of China in historical and future scenarios. Jabeen et al. (2017) found that the rise in maximum and minimum temperatures decreased the wheat yields using DSSAT and GIS across the Pothwar region of Pakistan. Araya et al. (2015) observed that simulated maize yield under future climate scenarios may increase slightly compared to historical period by 1.7% during the near future (2010–2039) for RCP4.5 and RCP8.5, subsequently 2.9% and 4.2%, and 3.5% and 3.8% during the middle (2040–2069) and end of the 21st century (2070–2099) for RCP4.5 and RCP8.5 in southwestern Ethiopia. Wilcox and Makowski (2014) summarized from 90 studies that the effects of higher CO_2 concentrations (>640ppm[①]) outweighed those of increasing temperature (up to +2°C) and moderate declines in precipitation (up to −20%), leading to augmenting yields. However, few studies have investigated the impact of single climate variables.

    The Huang-Huai-Hai Plain, which is located in Northern China, is the major grain-producing area in China (Yong et al., 2013). The conventional agricultural practice involves a rotation between winter wheat and summer maize, and this area provides approximately 70% of wheat and 30% of maize produced in China (Yang et al., 2015). Due to the extratropical monsoon climate, more than 70% of annual precipitation falls during summer (July to September) (Zhang et al., 2011). Thus, the winter wheat growing period has suffered from serious water deficits, and only 25%–40% of water demand is satisfied by rainfall (Mei et al., 2013). To maintain a high yield, the wheat is irrigated with pumped groundwater, which has led to environmental threats to society, such as

---

① 1ppm=10^{-6}。

groundwater lowering and surface subsidence (Zhang et al., 2005). Additionally, the warming temperatures and changing precipitation patterns caused by climate change have brought perturbations to regional crop production (Yu et al., 2014). Traditional climate-crop models based on statistical analyses have often produced conflicting conclusions because they use different elimination methods or consider different predictor variables (Shi et al., 2013). Therefore, it is of importance to capture the impacts of climate change as estimated using physically based crop models.

The objectives of this study were to understand the changes in climate variables during the winter wheat growing season over the Huang-Huai-Hai Plain from the last 30 years (1985–2014) to the next 30 years (2021–2050) under the RCP4.5 and RCP8.5 pathways, as well as the combined and isolated impacts of these variables on winter wheat yield.

5.2 Materials and methods

5.2.1 Study region

Study stations in this study were selected from agro-meteorological experimental stations where observations on wheat growth and development information were available. The selected stations had continuity in cultivar and to be disposed of detailed field management information for at least three years, and these stations were uniformly distributed. The 12 selected stations represent the general situation of the Huang-Huai-Hai Plain. The locations of these stations, as well as general information about them, are shown in Table 5-1. As we can see, annual mean precipitation and

Table 5-1 General information on the ten selected stations in the Huang-Huai-Hai Plain

Station	Longitude	Latitude	Location	Precipitation (mm)	Temperature (°C)	Solar radiation ($MJ \cdot m^{-2}$)
Tianjin	117°43′	39°03′	North	192.6	9.5	3760.8
Nangong	115°23′	37°22′	North	178.2	10.5	3922.1
Shangqiu	115°40′	34°27′	Central	315.1	11.5	3650.6
Yanzhou	116°51′	35°34′	Central	270.9	10.6	3893.9
Weifang	119°11′	36°45′	East	233.8	9.6	3997.1
Linyi	118°21′	35°03′	East	324.1	10.9	3993.6
Xinxiang	113°53′	35°19′	West	215.6	11.6	3769.3
Zhumadian	114° 01′	33°	West	470.0	12.3	3606.3
Xuzhou	117°29′	34°17′	South	369.7	11.8	3859.7
Shou Xian	116°47′	32°33′	South	509.6	12.3	3778.3

Notes: The values described in this table were average data for precipitation, temperature and solar radiation during the winter wheat growing period in 1981–2010 temperature values are systematically higher in the south than in the north, while solar radiation does not show obvious distribution patterns.

5.2.2 CERES-Wheat crop model

The CERES-Wheat model is a simulation system that predicts daily wheat growth, development and yield based on input information on the atmospheric and soil environment, cultivar factors and management (Ritchie et al., 1998). CERES-Wheat, as well as models representing other cereal crops included in DSSAT (e.g., CERES-Maize and CERES-Rice), have been widely used in optimizing the resource utilization and quantifying risks related to weather variations at scales ranging from fields to regions around the world (Timsina and Humphreys, 2006). Additionally, the applicability of CERES-Wheat has been tested in a wide range of field trials (Xiong et al., 2014). Thus, the CERES-Wheat model was selected to evaluate the impact of climate change in this study.

Weather, soil and management information

The minimal data sets required for model operation include weather information (daily global solar radiation, maximum and minimum temperatures, precipitation), soil information (classification and basic profile characteristics by soil layer) and management information (e.g., cultivar, planting, irrigation and fertilization information). In this study, the daily weather data from 1985–2014 collected at the ten stations were obtained from the National Meteorological Information Center of China. As the daily solar radiation data was not available, it was calculated with the Angstrom formula, which relates solar radiation to extraterrestrial radiation and relative sunshine duration (Martínez-Lozano et al., 1984). The projected daily weather data for 2021–2050 under RCP4.5 and RCP8.5 were derived from the BCC_CSM1.1 model and downscaled to the ten stations using the bilinear interpolation method (Yuan et al., 2012). Soil classification and profile characteristics were collected from the China Soil Scientific Database. In this study, we focus on evaluating the impact of climate change on wheat yield and identifying the contribution of each climate variable to the change. Thus, during the simulations, nitrogen stress was turned off, to avoid the impact of variable fertilizer levels among stations on the final result. On the other hand, water stress was turned on, and no irrigation was applied, in order to fully simulate the impact of climate change on water availability.

Model calibration and evaluation

In the CERES-Wheat model, there are seven main coefficients that control the development and growth of wheat (Ritchie et al., 1998). These coefficients must be calibrated and evaluated to meet the observed development and growth process under specific environmental conditions before further analyses (Hunt and Boote, 1998). In this study, the observed dates of the major stages of wheat development (i.e., the anthesis date and the maturity date) and the final yield, which were obtained from the agro-meteorological stations of the China Meteorological Administration, were used to calibrate and then evaluate the model at each station. General information on the 10 stations with chosen seasons and representative cultivar names used in model calibration and evaluation, as well as averaged data on growth stages and yields, are shown in Table 5-2.

Table 5-2 The average planting date, anthesis days (days after planting, ADAP), maturity days (days after planting, MDAP) and yield (HWAM) of the seasons selected for model calibration and evaluation

Station	Location	Season	Cultivar name	Planting date (MM.DD)	ADAP	MDAP	HWAM (kg·ha^{-1})
Tianjin	North	2006, 2007*, 2008	Beijing 9428	10.03	220	254	5300
Nangong	North	2002, 2003, 2004*, 2005	Han 6172	10.12	201	234	4586
Shangqiu	Central	2007, 2008, 2009*	Wenmai 6	10.15	191	225	5122
Yanzhou	Central	2006*, 2007, 2008, 2009	Jining 6	10.15	197	234	6301
Weifang	East	2007*, 2008, 2009	Jimai 22	10.08	210	241	6285
Linyi	East	2007*, 2008, 2009	Linmai 4	10.12	205	236	5271
Xinxiang	West	2004, 2005*, 2006, 2007	Xinmai 6	10.12	199	230	4285
Zhumadian	West	2006, 2007*, 2008, 2009	Zhengmai 9023	10.25	169	208	6299
Xuzhou	South	2005, 2006, 2007*, 2008	Xuzhou 24	10.17	193	228	3853
Shou Xian	South	2006, 2007, 2008, 2009*	Xumai 27	10.21	180	214	5460

Notes: The seasons marked with "*" were chosen for model evaluation

5.2.3 Simulated scenarios: past, future and isolated variables

During the wheat growing season, the average received solar radiation was 4.60% and 3.82% higher in 2021–2050 under the RCP4.5 and RCP8.5 pathways, respectively, when compared to 1985–2014 (Table 5-3). Negative changes were only found in Shou Xian, which is located in the west, with recorded reduction rates of 3.53% and 4.23%.

The average differences in the seasonal maximum temperature (TMAX) between 2021–2050 and 1985–2014 were 1.30°C (RCP4.5) and 1.28°C (RCP8.5), indicating that the maximum temperature during the wheat growing season becomes higher under the RCP4.5 scenario than under the RCP8.5. Specifically, in Nangong (north), Shangqiu (central), Yanzhou (central), Linyi (east) and Xinxiang (West), changes in TMAX were higher under RCP4.5 than under RCP8.5. Additionally, in the northern locations (Tianjin and Nangong), the increasing amplitude of TMAX was higher than that in other regions, with the average increase of 1.48°C and 1.45°C under RCP4.5 and RCP8.5, respectively.

Table 5-3 Simulation scenarios established to characterize the impact of climate change on winter wheat yields

Scenario	SRAD	TMAX	TMIN	RAIN	Purpose
SH	1985–2014	1985–2014	1985–2014	1985–2014	To simulate the yield level based on historical climate
SF	2021–2050	2021–2050	2021–2050	2021–2050	To simulate the yield level based on projected climates for both RCP4.5 and RCP8.5
SF-RAD	2021–2050	1985–2014	1985–2014	1985–2014	To simulate only the impact of radiation change on yield
SF-TEMP	1985–2014	2021–2050	2021–2050	1985–2014	Same as SH, but for maximum and minimum temperatures
SF-RAIN	1985–2014	1985–2014	1985–2014	2021–2050	Same as SH, but for rainfall

Increases in the minimum temperature (TMIN) during the wheat growing seasons were smaller than that in TMAX. The average changes in the TMIN and TMAX are 1.07°C and 1.09°C, respectively, which are higher under RCP8.5 than RCP4.5. Most stations experienced greater increases under the RCP8.5 scenario than under RCP4.5, except for Shangqiu, Xinxiang and Xuzhou. Similarly, the changes in TMIN were also higher in the north than in other regions, with increases of 1.49°C and 1.60°C. Precipitation (PRCP) increased by 15.01% under RCP4.5 and 16.44% under RCP8.5. The exceptions are Tianjin, Nangong and Shangqiu, for which PRCP shows a negative trend under RCP4.5 and/or RCP8.5.

To investigate the wheat yield response to climate change during the periods 1985–2014 and 2021–2050 and the relative contributions of changes in single climate variables, several simulation scenarios were set as shown in Table 5-3. The yield levels under historical climate conditions were simulated in SH using the observed daily maximum and minimum temperatures (TMAX and TMIN), precipitation (RAIN) and solar radiation (SRAD, estimated using the Angstrom formula) for 1985–2014 to strip the effects of irrigation and fertilization which can be different in different stations, while the predicted yields for future projected climate conditions were simulated in SF using TMAX, TMIN, RAIN and SRAD values for 2021–2050 under RCP4.5 and RCP8.5 respectively. To study the impact of change in each climate variable individually, simulations of SH-RAD, SF-TEMP and SF-RAIN were designed to simulate yields under historical climate conditions with only changes in SRAD, temperature (including TMAX and TMIN) and RAIN, respectively.

The impact of climate change on yield could be calculated by:

$$F_{ALL} = \frac{Y_{SF} - Y_{SH}}{Y_{SH}} \times 100\% \qquad \text{Formula 5-1}$$

$$F_{S} = \frac{Y_{SF-R_{AD}} - Y_{SH}}{Y_{SH}} \times 100\% \qquad \text{Formula 5-2}$$

$$F_{T} = \frac{Y_{SF-TEMP} - Y_{SH}}{Y_{SH}} \times 100\% \qquad \text{Formula 5-3}$$

$$F_{R} = \frac{Y_{SF-RAIN} - Y_{SH}}{Y_{SH}} \times 100\% \qquad \text{Formula 5-4}$$

where, F_{ALL}, F_S, F_T and F_R represent the impacts of changes in all climate variables, only SRAD, only temperature and only RAIN, respectively; Y_{SH} is the wheat yield that will be simulated based on historical climate; Y_{SF} is the wheat yield that will be simulated based on projected climates for both RCP 4.5 and RCP 8.5.

5.3 Results

5.3.1 Testing of CERES-Wheat model

The measured variables, specifically anthesis days after planting (ADAP), maturity days after planting (MDAP) and harvest weight at maturity (HWAM) at the selected agro-meteorological stations, were used to calibrate and evaluate the performance of the CERES-Wheat model for each selected station. The scatter plots and the normalized root mean square errors (NRMSEs) between the measured and simulated variables are shown in Figure 5-1 and Table 5-4.

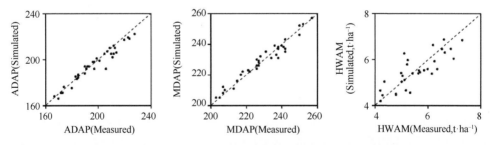

Figure 5-1 Comparison of simulated and measured anthesis days (ADAP, d), maturity days (MDAP, d) and yields (HWAM, kg·ha^{-1}) for selected ten stations. The dotted line is the 1:1 reference line

As described, the output of the calibrated model correlates well with the measured variables, especially during the growth stages. The ranges of NRMSEs for ADAP, MDAP and HWAM are between 0.8%–4.9%, 0.4%–2.5% and 2.4%–12.4%, respectively, and these ranges lie within acceptable limits (10% for growth duration and 15% for measured yields) (Ritchie et al., 1998).

Table 5-4 The calibrated genetic coefficients and the normalized root mean square errors (NRMSEs, %) values for ADAP (Anthesis, d), MDAP (Maturity, d) and HWAM (Yield, kg·ha^{-1})

Stations	Location	P1V (d)	P1D (%)	P5 (°C·d)	G1 (number·g^{-1})	G2 (mg)	G3 (g)	PHINT (°C·d)	NRMSE (%) ADAP (d)	MDAP (d)	HWAM (kg·ha^{-1})
Tianjin	North	34.2	52.3	652.5	28.6	22.6	1.6	95	2.5	0.5	2.4
Nangong	North	26.3	44.5	554.1	21.3	43.6	2.0	95	4.9	2.5	10.8
Shangqiu	Central	63.5	17.0	552.9	26.3	58.0	1.1	95	2.0	1.0	7.5
Yanzhou	Central	51.7	23.4	582.9	30.0	64.5	1.2	95	0.8	1.2	12.4
Weifang	East	6.8	66.6	473.0	26.2	38.2	2.0	95	0.8	0.4	5.9
Linyi	East	43.6	55.7	554.7	24.1	25.6	1.1	95	2.2	1.3	12.2
Xinxiang	West	42.0	48.3	498.2	29.8	62.9	1.9	95	0.8	1.3	8.6
Zhumadian	West	44.4	7.5	549.8	23.9	63.6	1.1	95	1.4	1.3	7.3
Xuzhou	South	64.5	27.6	642.7	20.0	56.1	1.1	95	1.3	1.3	11.4
Shou Xian	South	46.5	23.9	504.3	28.1	58.4	1.9	95	1.6	1.0	9.2

Notes: P1V is vernalization parameter (d), P1D is the photoperiod parameter (%), P5 is the grain filling duration parameter (°C·d), G1 is the grain parameter at anthesis (number·g^{-1}), G2 is the grain filling rate parameter (mg), G3 is the dry weight of a single stem and spike (g), PHINT is the interval between successive leaf tip appearances (°C·d)

5.3.2 Changes in growth duration and related climate variables

Changes in growth duration

In the model, the duration of growing stages was directly determined by temperature (Ritchie et al., 1998). In the context of global warming, a shortened growth period for wheat has been observed in both field observations and model predictions (Wang et al., 2008; Wei et al., 2012). In this study, based on the calibrated crop model, the ADAP (anthesis days after planting) and MDAP (maturity days after planting) were shortened in 2021–2050, compared with 1985–2014 (Figure 5-2). The dates of anthesis and maturity advanced in all selected stations within the Huang-Huai-Hai Plain. The average decrease in duration for ADAP was 8 and 9 days under RCP4.5 and RCP8.5, respectively, while the decrease in MDAP was 9 days for both pathways. Additionally, as depicted in Figure 5-2, although the length of ADAP and MDAP are truncated, the duration between anthesis and maturity shows no significant change as the ADAP decreased by the same amount as MDAP.

This result is identical to the findings based on long-term wheat phenology observations. Xiao et al. (2015) found that, while the phenological phases of winter wheat were shortening, the duration of the grain-filling stage even became slightly longer. It was believed that this phenomenon was a self-adaptation strategy which, in turn, not only prolonged growth stages but also enhanced the productivity of winter wheat. Correspondingly, our results show that the self-adaptation mechanism would also be effective under future scenarios. Additionally, they found that temperature during the grain-filling stage decreased as the grain-filling date advanced (Xiao et al., 2015, 2014), which also improved wheat production. In this study, we do not analyze the sensitivity of different wheat phenology to temperature warming, as we are more concerned with the overall tendency of climate conditions during the wheat growing season and the corresponding impact on wheat yields.

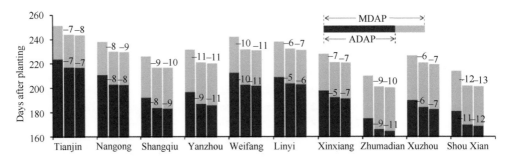

Figure 5-2 Changes in the number of days from planting to anthesis (ADAP, d) and from planting to maturity (MDAP, d). For each station, the columns from left to right represent the growing durations of 1985–2014 and 2021–2050 under the RCP4.5 scenario and 2021–2050 under the RCP8.5 scenario

Changes in climate variables

During the wheat growing season, the average received solar radiation was 4.60% and 3.82% higher in 2021–2050 under the RCP4.5 and RCP8.5 pathways, respectively, when compared to 1985–2014. Negative changes have been only found in Shou Xian, which is located in the west.

The average differences in the seasonal maximum temperature (TMAX) between 2021–2050 and 1985–2014 were 1.30°C (RCP4.5) and 1.28°C (RCP8.5), indicating that the maximum temperature during wheat growing seasons would become higher under the RCP4.5 pathway than in the RCP8.5 pathway. Specifically, in Nangong (north), Shangqiu (central), Yanzhou (central), Linyi (east) and Xinxiang (West), changes in TMAX were found to be higher under RCP4.5 than under RCP8.5. Additionally, in the northern locations (Tianjin and Nangong), the increasing amplitude of TMAX was higher than that of other regions, with the average increase of 1.48°C and 1.45°C under RCP4.5 and RCP8.5.

Increases in the minimum temperature (TMIN) during the wheat growing seasons resulted in a smaller effect than that in TMAX. The average changes in the TMIN were 1.07°C and 1.09°C, which are higher under RCP8.5 than RCP4.5. Most stations experienced greater increases under the RCP8.5 scenario than under RCP4.5, except for Shangqiu, Xinxiang and Xuzhou. The changes in TMIN were also higher in the north than in other regions, with increases of 1.49°C and 1.60°C (Table 5-5).

Table 5-5 Changes in seasonal solar radiation (SRAD), maximum temperature (TMAX), minimum temperature (TMIN) and rainfall accumulation (RAIN) during the winter wheat growing season between 1985–2014 and 2021–2050 under RCP4.5 and RCP8.5 scenario

Station	Change in SRDA (%)		Change in TMAX (°C)		Change in TMIN (°C)		Change in PRCP (%)	
	RCP4.5	RCP8.5	RCP4.5	RCP8.5	RCP4.5	RCP8.5	RCP4.5	RCP8.5
Tianjin	7.73	6.56	1.47	1.49	1.34	1.48	−4.68	6.31
Nangong	5.57	4.67	1.49	1.41	1.64	1.73	−3.87	11.97
Shangqiu	9.40	8.90	1.56	1.44	0.67	0.53	−0.90	-5.25
Yanzhou	1.05	0.33	1.12	1.05	1.28	1.33	14.26	16.48
Weifang	1.32	0.66	1.12	1.16	1.63	1.76	5.10	6.03
Linyi	4.05	3.43	1.15	1.09	0.50	0.58	15.70	19.45
Xinxiang	10.39	9.77	1.61	1.58	0.17	0.04	27.65	23.85
Zhumadian	4.80	3.59	1.08	1.10	1.07	1.11	33.97	29.13
Xuzhou	5.22	4.51	1.27	1.32	0.55	0.50	41.44	38.77
Shou Xian	−3.53	−4.23	1.09	1.12	1.81	1.86	21.42	17.63
Average	4.60	3.82	1.30	1.28	1.07	1.09	15.01	16.44

In terms of precipitation (PRCP), it increased by 15.01% under RCP4.5 and 16.44% under RCP8.5. The exceptions are Tianjin, Nangong and Shangqiu, for which PRCP shows a negative trend under RCP4.5 and/or RCP8.5. The increasing PRCP would inevitably relieve the water stress, which has been a serious problem in the study region.

5.3.3 Changes in yield and the contributions of single climate variables

Comparing the yield levels of simulation SF and SH, the climate changes from 1958–2014 to 2021–2050 positively impact winter wheat yields in the Huang-Huai-Hai Plain. The combined changes in SRAD, TMAX, TMIN and PRCP increased overall winter wheat yield by 17.37% and 15.76% under RCP4.5 and RCP8.5 in terms of the regional average, respectively (Table 5-6, F_{ALL}). Additionally, the percent increase was higher at the southern stations (Xuzhou and Shou Xian) than at the northern ones (Tianjin and Nangong).

Comparing the expected yield levels of simulation SF-RAD and SH (Figure 5-3B), radiation change (with other variables untouched), negatively impacts winter wheat yield in the study region. The change of SRAD in isolation reduced the yields by 4.49% and 5.4% under RCP4.5 and RCP8.5, respectively (Table 5-6, F_S). In addition, the negative impact was lower at the southern stations. It is conventionally assumed that increasing solar radiation is beneficial to photosynthesis and consequently yield.

This study nevertheless indicates that the expected benefit would be reversed by the subsequent higher water stress, and this mechanism will be discussed in detail in next section.

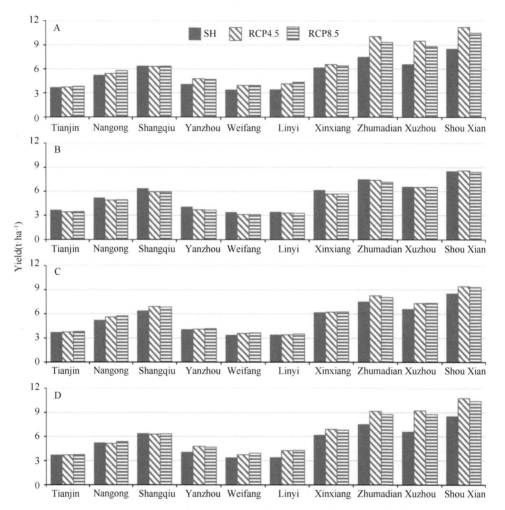

Figure 5-3 Comparisons of yields between simulation scenario SH and (A) SF; (B) SF-RAD; (C) SF-TEMP; and (D) SF-RAIN under RCP4.5 and RCP8.5 scenarios

Figure 5-3C depicts the changes in simulated yield caused by changes in temperature alone. It shows that warming temperatures have a positive impact on rainfed wheat yield in the Huang-Huai-Hai Plain. The overall rates of increase were 5.45% and 6.15% under RCP4.5 and RCP8.5, respectively (Table 5-6, F_T). Figure 5-3d depicts the changes in yield caused by changes in precipitation only. It shows that increasing precipitation during the wheat growing season is beneficial to wheat yields, except in Nangong and Shangqiu, where the precipitation decreases. This positive impact tends to be higher at southern stations (Table 5-6, F_R) due to the larger increase in precipitation at these stations. In addition, the relative contributions of precipitation were 14.79% (RCP4.5) and 14.16% (RCP8.5) which were much higher than the positive contributions from warming temperatures. Thus, the change in precipitation was the dominant climate variable that contributed to the positive impact of climate changes on wheat yield in the Huang-Huai-Hai Plain.

Table 5-6 The relative contributions of changes in each climate variable individually to yield changes for 2021–2050 under the RCP4.5 and RCP8.5 scenarios compared to 1985–2014

Station	F_{ALL}		F_S		F_T		F_R	
	RCP4.5	RCP8.5	RCP4.5	RCP8.5	RCP4.5	RCP8.5	RCP4.5	RCP8.5
Tianjin	0.33	3.48	−6.98	−6.10	0.98	2.67	0.01	1.51
Nangong	3.81	10.98	−5.67	−5.15	7.35	9.83	−1.51	2.81
Shangqiu	−1.01	−0.61	−6.53	−6.91	8.28	7.08	−1.53	−1.13
Yanzhou	17.06	15.59	−8.72	−9.49	0.57	2.45	16.92	14.75
Weifang	16.14	16.10	−8.32	−8.02	5.57	7.87	10.29	16.24
Linyi	21.64	27.34	−3.44	−4.04	−0.10	2.62	24.50	25.40
Xinxiang	6.52	3.87	−7.71	−7.90	0.88	1.48	11.56	10.15
Zhumadian	33.68	24.55	−1.41	−4.36	9.82	7.23	22.14	16.95
Xuzhou	43.99	33.52	−1.01	−0.82	10.93	11.13	39.51	33.00
Shou Xian	31.27	22.77	0.40	−1.20	10.22	9.17	25.99	21.89
Average	17.34	15.76	−4.94	−5.40	5.45	6.15	14.79	14.16

Notes: The unit is percent for each value

5.4 Discussion

5.4.1 Negative impact of increasing solar radiation

In this study, the solar radiation during the winter wheat growing seasons in 2021–2050 was 4.60% and 3.82% higher under the RCP4.5 and RCP8.5 scenarios respectively than those in 1985–2014. However, these increases had a negative impact on wheat yields (RECP 4.5: −4.49% and RCP8.5: −5.40%). The negative impacts were higher in the north than in the south. The result is inconsistent with previous studies, which suggest that crop yields are positively correlated with solar radiation, using both models (Wei et al., 2012) and statistical approaches (Tao et al., 2014).

It is generally recognized that increases in solar radiation would have a positive impact on wheat photosynthesis and consequently increase yield (Tao et al., 2008). However, by comparing the photosynthetic active radiation (PAR) conversion to the dry matter ratio before last leaf stage (PARUE) between simulations SH and SF-RAD, the results show that the PARUE is higher under conditions of low radiation, indicating that there is a decrease in the radiation conversion to dry matter ratio when the solar radiation increases under the RCP4.5 and RCP8.5 scenarios (Table 5-7).

Table 5-7 Changes in photosynthetically active radiation (PAR) conversion to dry matter ratio before the last leaf stage (PARUE) between simulation scenarios SH and SSF-RAD under the RCP4.5 and RCP8.5 scenarios

Station	PARUE for SH($g \cdot MJ^{-1}$)	Changes in PARUE (SF-RAD-SH, %)	
		RCP4.5	RCP8.5
Tianjin	0.89	−12.79	−12.02
Nangong	1.05	−12.03	−11.08
Shangqiu	1.16	−18.10	−18.10
Yanzhou	0.92	−11.64	−11.27
Weifang	0.78	−11.54	−12.39
Linyi	0.95	−9.12	−9.82
Xinxiang	1.06	−18.30	−16.72
Zhumadian	1.52	−9.19	−10.72
Xuzhou	1.14	−9.06	−8.16
Shou Xian	1.48	−4.05	−4.50
Average	1.10	−11.58	−11.48

In the CERES-Wheat model, PARUE is affected by nitrogen, temperature and water (Ritchie et al., 1998). In this study, all simulation scenarios were free of nitrogen stress; and the temperature and precipitation inputs were the same in simulations SH and SF-RAD. Thus, the nitrogen and temperature stress for photosynthesis were identical between simulations SH and SF-RAD. Therefore, the reduced PARUE is certainly caused by an increased water stress (increasing the soil evaporation in water balance). Comparing the relative moisture index (the ratio of difference between total evapotranspiration and precipitation to the total evapotranspiration) for SH and SF-RAD (Figure 5-4), we observe that the higher the relative moisture index is, the drier conditions are within a time period or a region. Figure 5-4 shows that the relative moisture index for SF-RAD is higher than that at SH. In other words, the increasing solar radiation has led to increased water stress, and consequently reduced PARUE and final yield. Additionally, as we can see, stations with lower relative moisture index values, such as Linyi, Zhumadian, Xuzhou, and Shou Xian, tend to experience much smaller scale of decrease in PARUE (Table 5-7) and yield (Table 5-6) that are much smaller. Thus, the impact of increases in radiation on winter wheat yield is closely linked with regional water situations. For stations in the southern parts of the with sufficient precipitation, the negative impact of increasing solar radiation is less apparent, whereas it is higher at the northern stations that experience serious precipitation shortages. In a conclusion, the supposed positive impact of increasing solar radiation is reversed by water deficit, as the radiation also induces higher atmospheric evapotranspiration demand.

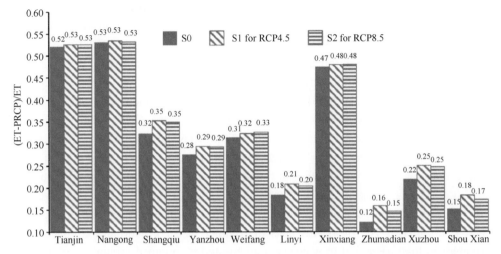

Figure 5-4 The characteristics in relative moisture index under historical scenario (S0), RCP4.5 scenario (S1) and RCP8.5 scenario (S2). The relative moisture index was estimated as the ratio of difference between/ of total evapotranspiration and precipitation to the total evapotrans-piration

5.4.2 Positive impact of warming temperature and increasing precipitation

The impact of temperature (SF-TEMP) or precipitation (SF-RAIN) increase during the winter wheat growth period on wheat yield is positive. This positive impact of warming on wheat yield in the Huang-Huai-Hai Plain differs from the results of studies for China (Wang et al., 2008; You et al., 2009), Africa (Wolfram and David, 2010), South Asia (Jerry et al., 2012), and the would (David and Christopher 2007; Lobell et al., 2011). However, it is supported by the results of Tao et al. (2008, 2014) and Xiao et al. (2008), who concluded that the wheat yield had benefited from climate warming in northern and northwest China. The impact of warming on crop yields is spatially variable in our study. Generally, a warming trend reduces crop productivity in areas at low latitudes where optimal condition of temperatures already exist. However it may cause beneficial effects at high latitudes (Wei et al., 2012). Tao et al. (2014)concluded that increasing temperatures increased wheat yield in Northern China but reduced wheat yield in Southern China, with the explanation that the temperature in Northern China was lower than the optimum temperature for wheat. Additionally, although the period of vegetative growth (VGP) and the over growing period tend to be shorter, the length of the reproductive growth stage (RGP, anthesis-maturity) shows no significant change. It indicates that the wheat cultivars have has a high thermal requirement during the RGP. Grain filling rates became higher as temperature increases until reaching the optimum threshold; and grain yield would consequently increase. The self-adaptability of wheat cultivars in the last ten years (2001–2009) has compensated for the impact of climate change.

Rainfall has always been a limiting factor in agricultural production in China (Piao et al., 2010), especially in the Huang-Huai-Hai Plain, where only 25%–40% of water demand is satisfied by rainfall during the wheat growing seasons (Mei et al., 2013). Despite the uncertainty in precipitation amounts and the spatial patterns simulated by climate models (Xin et al., 2013), our study showed that the increasing rainfall in the next 30 years could provide important benefits in terms of the wheat yield. Many other studies have investigated the relative contribution of changing rainfall to the increase of wheat yield in China during the past several decades, based on long-term field observations or statistical datasets, and found that there was no significant influence of rainfall on yield (Tao et al., 2014; Wei et al., 2012; Xiao and Tao, 2014). One reason for this result is that rainfall did not change substantially during the studied period (Wei et al., 2012). The other reason is that winter wheat in Northern China is well irrigated, so that the deficit in water is compensated by pumped groundwater, and the impact of rainfall changes was smoothed over. However, in our study, to fully reenact the rainfall changes to yield, the wheat growth and development was simulated under rainfed conditions. Thus, the contributions of increasing rainfall to wheat yield were higher than those in other studies.

5.5 Conclusions

Climate change during the next 30 years (2021–2050) was found to have a positive impact on simulated wheat yield for two Representative Concentration Pathways. The largest contribution is caused by increasing precipitation during the growing season, which is inconsistent with the conclusions based on statistical models of observed data. In those previous studies, the wheat was irrigated; thus, the impact of changes in precipitation is weakened. Interestingly, the increasing solar radiation shows a negative impact on wheat yield in our study. We further found that this reverse was caused by the higher evapotranspiration that resulted from increases in solar radiation, which consequently increases water stress. Most other studies have shown that crop yield is positively correlated with solar radiation, mainly because the wheat was irrigated in their simulations, Thus, the water stress remained at low levels, even though the evapotranspiration was increased by solar radiation.

References

Araya A, Hoogenboom G, Luedeling E, Hadgu K M, Kisekka I, Martorano L G. 2015. Assessment of maize growth and yield using crop models under present and future climate in Southwestern Ethiopia. Agricultural and Forest Meteorology, 214-215: 252-265.

David B L, Christopher B F. 2007. Global scale climate-crop yield relationships and the impacts of recent warming. Environmental Research Letters, 2(1): 014002.

Hunt L A, Boote K J. 1998. Data for model operation, calibration, and evaluation. *In*: Tsuji G Y, Hoogenboom G, Thornton P K. Understanding Options for Agricultural Production. Dordrecht, Netherlands: Springer: 9-39.

Jabeen M, Gabriel H F, Ahmed M, Mahboob M A, Iqbal J. 2017. Studying Impact of Climate Change on Wheat Yield by Using DSSAT and GIS: A Case Study of Pothwar Region, Quantification of Climate Variability, Adaptation and Mitigation for Agricultural Sustainability. New York: Springer International Publishing: 387-411.

Jerry K, Tim H, Andre D, Tim W. 2012. Climate change impacts on crop productivity in Africa and South Asia. Environmental Research Letters, 7(3): 034032.

Jones J W, Tsuji G Y, Hoogenboom G, Hunt L A, Thornton P K, Wilkens P W, Imamura D T, Bowen W T, Singh U. 1998. Decision support system for agrotechnology transfer: DSSAT v3. *In*: Tsuji G Y, Hoogenboom G, Thornton P K. Understanding Options for Agricultural Production. Dordrecht, Netherlands: Springer: 157-177.

Ju H, van der Velde M, Lin E, Xiong W, Li Y. 2013. The impacts of climate change on agricultural production systems in China. Climatic Change, 120(1-2): 313-324.

Keating B A, Carberry P S, Hammer G L, Probert M E, Robertson M J, Holzworth D, Huth N I, Hargreaves J N G, Meinke H, Hochman Z, McLean G, Verburg K, Snow V, Dimes J P, Silburn M, Wang E, Brown S, Bristow K L, Asseng S, Chapman S, McCown R L, Freebairn D M, Smith C J. 2003. An overview of APSIM, a model designed for farming systems simulation. European Journal of Agronomy, 18(3-4): 267-288.

Lobell D B, Schlenker W, Costa-Roberts J. 2011. Climate trends and global crop production since 1980. Science, 333(6042): 616-620.

Martínez-Lozano J A, Tena F, Onrubia J E, De La Rubia J. 1984. The historical evolution of the Ångström formula and its modifications: Review and bibliography. Agricultural and Forest Meteorology, 33(2-3): 109-128.

Mei X, Kang S, Yu Q, Huang Y, Zhong X, Gong D, Huo Z, Liu E. 2013. Pathways to synchronously improving crop productivity and field water use efficiency in the North China Plain. Scientia Agricultura Sinica, 46(6): 1149-1157.

Piao S, Ciais P, Huang Y, Shen Z, Peng S, Li J, Zhou L, Liu H, Ma Y, Ding Y, Friedlingstein P, Liu C, Tan K, Yu Y, Zhang T, Fang J. 2010. The impacts of climate change on water resources and agriculture in China. Nature, 467(7311): 43-51.

Ritchie J T, Singh U, Godwin D C, Bowen W T. 1998. Cereal growth, development and yield. *In*: Tsuji G Y, Hoogenboom G, Thornton P K. Understanding Options for Agricultural Production. Dordrecht, Netherlands: Springer: 79-98.

Schlenker W, Roberts M J. 2009. Nonlinear temperature effects indicate severe damages to US crop yields under climate change. Proceedings of the National Academy of Sciences of the United States of America, 106(37): 15594-15598.

Shi W, Tao F, Zhang Z. 2013. A review on statistical models for identifying climate contributions to crop yields. Journal of Geographical Sciences, 23(3): 567-576.

Tao F, Yokozawa M, Liu J, Zhang Z. 2008. Climate-crop yield relationships at provincial scales in China and the impacts of recent climate trends. Climate Research, 38(1): 83-94.

Tao F, Zhang Z, Xiao D, Zhang S, Rötter R P, Shi W, Liu Y, Wang M, Liu F, Zhang H. 2014. Responses of wheat growth and yield to climate change in different climate zones of China, 1981–2009. Agricultural and Forest Meteorology, 189-190: 91-104.

Timsina J, Humphreys E. 2006. Performance of CERES-Rice and CERES-Wheat models in rice–wheat systems: A review. Agricultural Systems, 90(1-3): 5-31.

Wang H L, Gan Y T, Wang R Y, Niu J Y, Zhao H, Yang Q G, Li G C. 2008. Phenological trends in winter wheat and spring cotton in response to climate changes in northwest China. Agricultural and Forest Meteorology, 148(8-9): 1242-1251.

Wei X, Ian H, Erda L, Declan C, Yue L, Wenbin W. 2012. Untangling relative contributions of recent climate and CO_2 trends to national cereal production in China. Environmental Research Letters, 7(4): 044014.

Wilcox J, Makowski D. 2014. A meta-analysis of the predicted effects of climate change on wheat yields using simulation studies. Field Crops Research, 156: 180-190.

Williams J R, Renard K G, Dyke P T. 1983. EPIC: A new method for assessing erosion's effect on soil productivity. Journal of Soil and Water Conservation, 38(5): 381-383.

Wolfram S, David B L. 2010. Robust negative impacts of climate change on African agriculture. Environmental Research Letters, 5(1): 014010.

Xiao D, Moiwo J P, Tao F, Yang Y, Shen Y, Xu Q, Liu J, Zhang H, Liu F. 2015. Spatiotemporal variability of winter wheat phenology in response to weather and climate variability in China. Mitigation and Adaptation Strategies for Global Change, 20(7): 1191-1202.

Xiao D, Tao F, Shen Y, Liu J, Wang R. 2014. Sensitivity of response of winter wheat to climate change in the North China Plain in the last three decades. Chinese Journal of Eco-Agriculture, 22(4): 430-438.

Xiao D, Tao F. 2014. Contributions of cultivars, management and climate change to winter wheat yield in the North China Plain in the past three decades. European Journal of Agronomy, 52(Part B): 112-122.

Xiao G, Zhang Q, Yao Y, Zhao H, Wang R, Bai H, Zhang F. 2008. Impact of recent climatic change on the yield of winter wheat at low and high altitudes in semi-arid northwestern China. Agriculture, Ecosystems & Environment, 127(1-2): 37-42.

Xin X G, Wu T W, Li J L, Zai Zhi W, Wei Ping L, Fang Hua W. 2013. How well does BCC_CSM1.1 reproduce the 20th century climate change over China? Atmospheric and Oceanic Science Letters: 6(1): 21-26.

Xiong W, van der Velde M, Holman I P, Balkovic J, Lin E, Skalský R, Porter C, Jones J, Khabarov N, Obersteiner M. 2014. Can climate-smart agriculture reverse the recent slowing of rice yield growth in China? Agriculture, Ecosystems & Environment, 196: 125-136.

Yang J Y, Mei X R, Huo Z G, Yan C R, Ju H, Zhao F H, Liu Q. 2015. Water consumption in summer maize and winter wheat cropping system based on SEBAL model in Huang-Huai-Hai Plain, China. Journal of Integrative Agriculture, 14(10): 2065-2076.

Yong B, Ren L, Hong Y, Gourley J J, Chen X, Dong J, Wang W, Shen Y, Hardy J. 2013. Spatial–temporal changes of water resources in a typical semiarid basin of North China over the past 50 years and assessment of possible natural and socioeconomic causes. Journal of

Hydrometeorology, 14(4): 1009-1034.

You L, Rosegrant M W, Wood S, Sun D. 2009. Impact of growing season temperature on wheat productivity in China. Agricultural and Forest Meteorology, 149(6-7): 1009-1014.

Yu Q, Li L, Luo Q, Eamus D, Xu S, Chen C, Wang E, Liu J, Nielsen D C. 2014. Year patterns of climate impact on wheat yields. International Journal of Climatology, 34(2): 518-528.

Yuan B, Guo J, Ye M, Zhao J. 2012. Variety distribution pattern and climatic potential productivity of spring maize in Northeast China under climate change. Chinese Science Bulletin, 57(26): 3497-3508.

Yuan W, Cai W, Chen Y, Liu S, Dong W, Zhang H, Yu G, Chen Z, He H, Guo W, Liu D, Liu S, Xiang W, Xie Z, Zhao Z, Zhou G. 2016. Severe summer heatwave and drought strongly reduced carbon uptake in Southern China. Scientific Reports, 6(1): 18813.

Zhang X, Chen S, Liu M, Pei D, Sun H. 2005. Improved water use efficiency associated with cultivars and agronomic management in the North China Plain. Agronomy Journal, 97(3): 783-790.

Zhang X, Chen S, Sun H, Shao L, Wang Y. 2011. Changes in evapotranspiration over irrigated winter wheat and maize in North China Plain over three decades. Agricultural Water Management, 98(6): 1097-1104.

Chapter 6　The impacts of climate change on wheat yield based on the DSSAT-CERES-Wheat model under the RCP8.5 scenario in the Huang-Huai-Hai Plain, China

Abstract

Progress in understanding the impact of climate change on crop yields is essential for agricultural climate adaptation, especially for the Huang-Huai-Hai Plain, an area that is particularly vulnerable to global warming. The Huang-Huai-Hai Plain is part of the mid-high latitude region. In this study, we analyzed changes in temperature, solar radiation, and precipitation during the winter wheat growing season, comparing their performance in the baseline period (1981–2010) to short-term (2010–2039), medium-term (2040–2069), and long-term (2070–2099) projections under the RCP8.5 scenario. The relative impact of these changes on winter wheat yield was revealed with the CERES-Wheat model. Results showed that maximum and minimum temperatures (TMAX and TMIN), solar radiation (SRAD), and precipitation (PRCP) during the wheat season increased under the RCP8.5 scenario, and when the fertilizer efficiency of CO_2 was not considered, changes in these variables increased wheat yield by 13.1% (short-term), 14.8% (medium-term), and 29.1% (long-term). Increasing precipitation increased wheat yield by 11.7% (short-term), 16.3% (medium-term), and 27.7% (long-term), and thus precipitation contributed most to the total climate change impact. However, warming temperatures decreased wheat yield by 3.6% (short-term), 5.8% (medium-term), and 5.8% (long-term), with a positive impact in the south and a negative impact in the north. Our analysis demonstrated that in the Huang-Huai-Hai Plain, advantageous increases in thermal resources were counteracted by the disadvantageous, aggravated water deficits caused by warming temperatures.

6.1 Introduction

Agriculture is one of the most sensitive systems to climate change because it is directly exposed to the aerial environment (Piao et al., 2010). Global warming increases the thermal resources for high latitudes, and expands arable land (Ju et al., 2013), while exacerbating heat and drought stress for arid and Semi-arid regions with fluctuating food supplies (Yuan et al., 2016). Thus, studying the impact of climate change, and how to benefit from positive impacts and avoid negative impacts are of great importance to agricultural adaptation and national food security (Ju et al., 2013; Piao et al., 2010).

Statistical models and numerical simulations have been widely used to detect climate-crop relationships (Shi et al., 2013). Based on long term yield data, a time-series regression model at the station level between yield series and series of several climate variables can be established. In this model, the regression coefficients represent the negative or positive impact of climate change (David and Christopher, 2007; Tao et al., 2008). In recent years, the statistical models that rely on information from multiple stations, namely panel models, were documented to be better at predicting crop responses to temperature change (Schlenker and Roberts, 2009; Wolfram and David, 2010; Tao et al., 2014). However, the influence of non-climatic factors (such as cultivar and fertilizer change) on yield fluctuations needs to be eliminated. Although some elimination methods, such as the first-difference yield and the linear detrended yield, have been widely used in several studies, their consistency remains questionable. For example, the linear detrended method is based on the notion that crop management and variety renewal changed linearly with time which does not match the reality (Xiong et al., 2014). Thus, the reality of these results remain questionable and the opposite impact has been documented (Shi et al., 2013).

In recognizing the complex interactions between crop growth and environmental factors, numerical simulation models have become a popular research tool for agro-meteorological researchers in recent years. Dynamic crop models, such as the Erosion Productivity Impact Calculator (EPIC) model (Williams et al., 1983), APSIM (Keating et al., 2003), and the Decision Support System for Agrotechnology Transfer (DSSAT) (Jones et al., 1998), have been tested and used to quantify water, nitrogen, and weather responses at field or regional scales around the world. Of these models, DSSAT contains models for different crops and can quantitatively predict the growth and production of annual field crops with the interactions of aerial and soil environments, cultivar factors, and management (Ritchie et al., 1998). DSSAT has been used to assess the impacts of climate change and climate extremes for rice, wheat, and maize in different regions of China for historical and future scenarios. However, few studies have investigated the impact of changes in a single climate variable in isolation.

Traditional climate-crop relationships based on statistical analyses often produced conflicting conclusions because they used different elimination methods or considered different predictor variables (Shi et al., 2013). Consequently, it is necessary to understand the impact of climate change using physically based crop models. The objectives of this study were to understand the shifts of climate variables during the winter wheat growing season in the Huang-Huai-Hai Plain under the RCP8.5 scenario, and to determine the relative impact of changes in each variable on winter wheat yield in isolation.

6.2 Materials and methods

6.2.1 Study region

The Huang-Huai-Hai Plain contains the major grain producing areas in China (Yong et al., 2013). The main crop pattern is winter wheat and summer maize rotation system, which provides about 70% and 30% of wheat and maize production in China, respectively (Yang et al., 2015). Due to the extratropical monsoon climate, more than 70% of annual precipitation falls during summer seasons (July to September) (Zhang et al., 2011). Thus, the winter wheat growing period experiences a serious water deficit with only 25%–40% of water demand satisfied by rainfall (Mei et al., 2013). To maintain a high yield, wheat is irritated with pumped groundwater which has led to environmental threats, such as groundwater lowering and surface subsidence (Zhang et al., 2005). Additionally, the warming temperature and changing precipitation pattern have induced perturbations to regional crop production (Yu et al., 2014).The Huang-Huai-Hai Plain can be divided into six sub-regions in terms of climate conditions and agricultural management practices, as shown in Figure 6-1. Detailed information of the six sub-regions is described by Li et al. (2015).

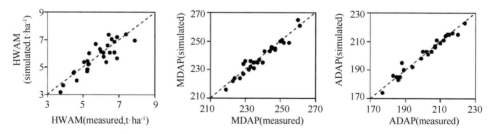

Figure 6-1 Comparison between measured and simulated values of ADAP (anthesis days after planting), MDAP (maturity days after planting), and HWAM (harvest yield at maturity). The dotted line is the 1:1 reference line

6.2.2 CERES-Wheat model

The CERES-Wheat model is a simulation system that predicts daily wheat growth, development, and yield based on the input information of aerial and soil environment, cultivar factors, and management (Ritchie et al., 1998). CERES-Wheat, along with other cereal crops included in DSSAT such as CERES-Maize and CERES-Rice, has been widely used in optimizing the use of resources and quantifying risk related to weather variations at field or regional scales around the world (Timsina and Humphreys, 2006). Additionally, the applicability of CERES-Wheat has been tested over a wide range of field trials (Xiong et al., 2014).

Weather, soil, and management information. Input data sets for model operation include weather information (daily solar radiation, maximum and minimum temperatures, precipitation), soil information (classification and basic profile characteristics by soil layer), and management information (e.g. cultivar, planting, irrigation, and fertilization information). The $0.5° \times 0.5°$ daily weather data under the RCP8.5 scenario were derived from the HadGEM2-ES model, and divided into four groups respectively for the baseline period (1981–2010), short-term period (2010–2039), medium-term period (2040–2069), and long-term period (2070–2099). The $0.5° \times 0.5°$ soil classification and profile characteristics were collected from the Harmonized World Soil Database (HWSD). In this study, no nitrogen stress was included, but water stress was, and no irrigation was applied in order to fully simulate the impact of climate change on water availability. As for the planting date for each grid and each year, last day over of 15°C of each year was considered a suitable sowing date, which was counted by the method of 5 days gliding average.

Model calibration and evaluation. In the CERES-Wheat model, coefficients that control the development and growth of wheat (Ritchie et al., 1998) must be calibrated and evaluated under specific environmental conditions before being used to analyze the impact of climate (Hunt and Boote, 1998). In this study, we selected one of the most popular wheat cultivars as representative. The actual anthesis date, maturity date, and yield data used for calibrating the cultivar coefficients were obtained from the agro-meteorological stations of the China Meteorological Administration (Table 6-1). General information on the six stations with chosen seasons and representative cultivar name used for model calibration and evaluation, and the average values of growth stages and yields are listed in Table 6-1.

6.2.3 Simulation design

To investigate the response of wheat yield to changes in each climate variable between baseline and the RCP8.5 scenario, five simulation scenarios were set as shown in Table 6-2. Scenario S0 was simulated the wheat yield under the baseline climate, which represents the baseline daily maximum and minimum temperatures (TMAX and TMIN), precipitation (RAIN), and solar radiation (SRAD); Scenarios S1, S2, S3, and S4 were

Table 6-1 The average planting date, days to anthesis (days after planting, ADAP), days to maturity (days after planting, MDAP), and yield (HWAM) of the seasons selected for model calibration and evaluation

Station	Sub-region	Season	Cultivar	Planting date (MM.DD)	ADAP	MDAP	HWAM (kg·ha^{-1})
Shenzhou	I	2003, 2004, 2005, 2007, 2009*	SX733	10.07	210	244	4285
Baodi	II	1996*, 1997*, 2000, 2001, 2002, 2004	JD8	10.03	220	254	5300
Huimin	III	2004, 2005, 2006, 2007, 2008*	LM23	10.12	201	234	4586
Zibo	IV	2002*, 2005, 2006	JM20	10.12	205	236	5271
Suzhou	V	2001, 2002, 2003, 2004*	WM19	10.15	191	225	5122
Huaian	VI	2001*, 2002*, 2004, 2005, 2006, 2007	WM6	10.21	180	214	5460

Notes: The seasons with "*" were chosen for model evaluation

simulated the wheat yield based on one or more climatic variable of RCP8.5 (temperature[T], T+SRAD, T+SRAD+PRCP, T+SRAD+PRCP+CO$_2$) with the remaining variables held constant.

Table 6-2 Simulation scenarios set to identify the impact of climate changes on winter wheat yield

Scenario	TMAX	TMIN	SRAD	PRCP	Purpose
S0	Baseline	Baseline	Baseline	Baseline	Simulate yield under baseline climate conditions
S1	RCP8.5	RCP8.5	Baseline	Baseline	Simulate yield under RCP8.5 temperatures, with SRAD and PRCP held constant
S2	RCP8.5	RCP8.5	RCP8.5	Baseline	Simulate yield under RCP8.5 temperatures and SRAD, with PRCP held constant
S3	RCP8.5	RCP8.5	RCP8.5	RCP8.5	Simulate yield under RCP8.5 climate conditions
S4	RCP8.5	RCP8.5	RCP8.5	RCP8.5	Simulate yield with the fertilizer efficiency of CO$_2$. Baseline period, 380 ppm; short-term, 423 ppm; medium-term, 571 ppm; long-term, 798 ppm.

The impact of climate change on yield was calculated by:

$$F_T = \frac{Y_{S1} - Y_{S0}}{Y_{S0}} \times 100\% \qquad \text{Formula 6-1}$$

$$F_S = \frac{Y_{S2} - Y_{S1}}{Y_{S0}} \times 100\% \qquad \text{Formula 6-2}$$

$$F_P = \frac{Y_{S3} - Y_{S2}}{Y_{S0}} \times 100\% \qquad \text{Formula 6-3}$$

$$F_{ALL} = \frac{Y_{S3} - Y_{S0}}{Y_{S0}} \times 100\% \qquad \text{Formula 6-4}$$

$$F_C = \frac{Y_{S4} - Y_{S3}}{Y_{S0}} \times 100\% \qquad \text{Formula 6-5}$$

$$F_{ALL-C} = \frac{Y_{S4} - Y_{S0}}{Y_{S0}} \times 100\% \qquad \text{Formula 6-6}$$

where, F_T, F_S, F_P, and F_C represent the relative impact of temperature changes, SRAD changes, precipitation changes, and the fertilizer efficiency of elevated CO_2, respectively; and F_{ALL} and F_{ALL-C} represent the total impact of climate change without and with fertilizer efficiency of elevated CO_2.

6.3 Results

6.3.1 Model calibration and validation

The coefficients of the chosen variety that represents the development and growth of wheat were calibrated and validated using at least three seasons of actual data (Table 6-1). Values of coefficients for wheat are shown in Table 6-3. The three coefficients (P1V, P1D, and P5) that affect the phenological development were estimated using the observed phenological dates (anthesis and maturity date). G1, G2, and G3 determine the grain number, grain weight, and the spike number, respectively. However, due to the lack of yield analysis data, these coefficients were estimated using harvest yield data only. PHINT is the thermal time between successive tip leaf appearances. The normalized root mean square error (NRMSE) listed in Table 6-3 represents the error between the simulated anthesis date (ADAP), maturity date (MDAP), and yield (HWAM) based on the observations.

Overall, the simulated phenological development was consistent with the observations (Figure 6-1). The NRMSE for anthesis date ranges from 0.3% to 1.3%, and it ranges from 0.8% to 1.5% for the maturity date (Table 6-3). Due to the lack of

Table 6-3 Genetic coefficients and the normalized root mean square error (NRMSE, %) of ADAP, MDAP and HWAM

Station	Location	P1V	P1D	P5	G1	G2	G3	PHINT	NRMSE (%)		
									ADAP	MDAP	HWAM
Shenzhou	I	19.6	38.8	557.0	30.0	65.0	1.9	95.0	0.9	1.1	6.8
Baodi	II	13.2	63.3	634.1	17.2	46.3	1.4	95.0	1.2	1.3	10.7
Huiming	III	13.7	57.9	560.2	29.0	64.2	1.1	95.0	0.9	0.8	11.4
Zibo	IV	9.1	79.0	602.4	20.9	43.8	1.0	95.0	0.3	0.9	11.5
Suzhou	V	15.0	63.0	583.1	15.4	57.4	1.2	95.0	1.3	1.5	8.9
Huaian	VI	10.1	62.0	685.3	16.0	50.5	1.1	95.0	1.9	1.4	6.0

P1V: Days required for vernalization (optimum vernalizing temperature); P1D: Photoperiod response (% reduction in rate/10h drop in photoperiod); P5: Grain filling phase duration (°C·d); G1: Kernel number per unit canopy weight at anthesis (#/g); G2: Standard kernel size under optimum conditions (mg); G3: Standard non-stressed mature tiller weight (including grain) (g[dry weight]); PHINT: Interval between successive leaf tip appearances (°C·d) yield data for calibration, the NRMSE values of yield are higher than the phenological development, ranging from 6.0% to 11.5%, which are still under the acceptance level of 15%. This suggests that the calibrated coefficients can be applied for the purpose of the present study.

6.3.2 Simulated changes of the phenological phase

In the global background of climate warming, historic results from either field observations or numerical models demonstrated that climate warming had shortened the length of the wheat-growing period (Wang et al., 2008). Based on the calibrated CERES-Wheat model, the length of the growth period under the RCP8.5 scenario was simulated over the short-term (2010–2039), medium-term (2040–2069), and long-term (2070–2099) (Figure 6-2). For the short, medium, and long-term the length from planting to maturity (MDAP) was shortened by 4 days, 15 days, and 25 days (averaged over all sub-regions), respectively, while the length from planting to anthesis (ADAP) was shortened by 4 days, 15 days, and 24 days (averaged over all sub-regions), respectively.

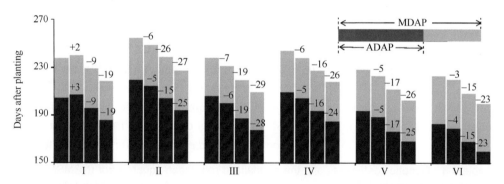

Figure 6-2 Changes in the number of days from planting to anthesis (ADAP) and planting to maturity (MDAP) under the RCP8.5 scenario compared to the baseline period. For each sub-region, columns from left to right represent the growing duration of baseline, and those over the short-term (2010–2039), medium-term (2040–2069), and long-term (2070–2099)

Interestingly, the length of the growth period between anthesis and maturity, which is important for grain filling and grain weight, did not change significantly. Field studies of wheat phenology also showed the same result. Wang et al. (2008) found that the growth period from anthesis to milk was prolonged by 8.2 days during 1981–2004 in northwest China. Early anthesis and maturity lead to a lower temperature during the grain-filling stage in the North China Plain, which not only prolongs growth stages but also enhances productivity of winter wheat. Xiao et al. (2015) called this the self-adaptation of winter wheat. Our result indicates that this self-adaptation also enables grain-filling to keep with enough length.

6.3.3 Changes of climatic variables during the wheat-growing period

Figure 6-3 shows the differences in maximum temperature (TMAX, Figure 6-3A), minimum temperature (TMIN, Figure 6-3B), solar radiation (SRAD, Figure 6-3C), and precipitation accumulation (PRCP, Figure 6-3D) during the winter wheat growing season between simulations under the RCP8.5 scenario (short-term, medium-term, and long-term) and the field observations during the baseline period. The TMAX and

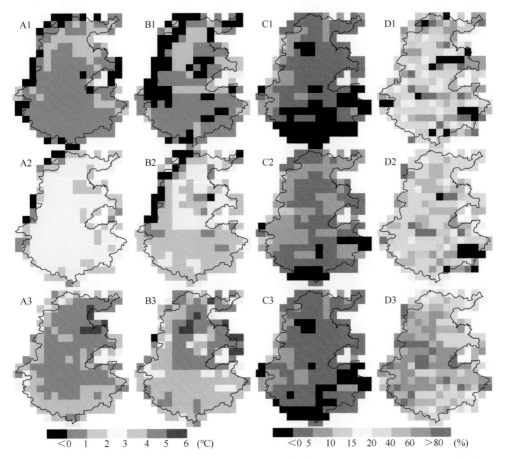

Figure 6-3 Changes in maximum temperature (A), minimum temperature (B), solar radiation (C) and precipitation (D) of the wheat-growing period for 1: short-term (2010–2039); 2: medium-term (2040–2069); and 3: long-term (2070–2099) compared to the baseline period

TMIN of most grids are expected to increase except in the mountain areas of the northwest region. The average increases of TMAX (TMIN) are 0.55°C (0.31°C), 2.17°C (1.62°C), and 3.86°C (3.56°C) for the short-term, medium-term, and long-term, respectively. Likewise, the seasonal SRAD and PRCP are also expected to increase. The expected increases that of PRCP (16.73% [short-term], 28.64% [medium-term], and 50.09% [long-term]) are much higher than those of SRAD (1.69% [short-term], 4.24% [medium-term], and 2.36% [long-term]) (Table 6-4). However, the SRAD of north areas in the study region has a decreasing trend, especially for the short-term.

Table 6-4 Changes in the maximum temperature (TMAX), minimum temperature (TMIN), solar radiation (SRAD), and precipitation accumulation (PRCP) during the winter wheat growing season between RCP8.5 scenario and field observations during the baseline period

Sub-region	Changes in TMAX (°C)			Changes in TMIN (°C)			Changes in SRDA (%)			Changes in PRCP (%)		
	S	M	L	S	M	L	S	M	L	S	M	L
I	0.11	1.51	3.02	−0.72	0.54	2.47	5.37	6.25	4.16	18.03	36.69	58.59
II	0.86	2.47	4.24	0.55	1.97	4.07	3.73	5.09	3.78	10.07	30.62	31.06
III	0.83	2.68	4.36	0.71	2.24	4.23	2.19	4.25	2.20	16.54	32.15	58.80
IV	0.41	2.17	4.05	0.79	2.13	4.05	2.00	4.48	2.53	22.32	36.23	57.94
V	0.63	2.26	3.96	0.38	1.62	3.52	0.85	4.26	2.10	18.84	23.54	55.06
VI	0.41	2.00	3.68	0.44	1.60	3.44	−3.73	0.78	−0.63	13.66	18.19	34.77
Average	0.55	2.17	3.86	0.31	1.62	3.56	1.69	4.24	2.36	16.73	28.64	50.09

6.3.4 Impacts of different climate variables on wheat yield

Wheat yields were simulated using one or more climatic variables of RCP8.5 (temperature, temperature + radiation, temperature + radiation + precipitate) with the other variables remaining at the level of the baseline period. The relative changes in yield were calculated using formulas (1) to (4). Figure 6-4 shows the relative impact of increasing temperature (A), radiation (B), precipitation (C), and the combined effect of these variables (D). The impact of warming temperature differs between the southern and northern Huang-Huai-Hai Plain (Figure 6-4A). The impact in southern sub-regions (V and VI) was positive, but was negative in the northern sub-regions (Table 6-5). On average, the increasing temperature would decrease wheat yield by 3.6% (short-term), 5.8% (medium-term), and 5.8% (long-term).

Changes in solar radiation and precipitation under the RCP8.5 scenario would positively impact wheat yield (Figure 6-4B and C). The increasing solar radiation would elevate wheat yield by 5.0% (short-term), 4.3% (medium-term), and 7.1% (long-term); increasing precipitation would elevate wheat yield by 11.7% (short-term), 16.3% (medium-term) and 27.7% (long-term) (Table 6-5). However, the impact of precipitation was negative in the southern part of the study region (Figure 6-4C), mainly because the increasing precipitation would raise risks of logging damage.

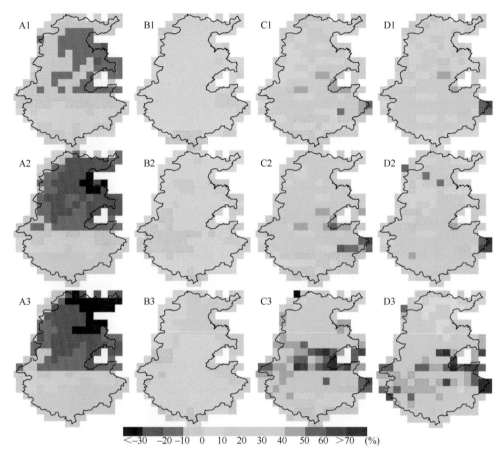

Figure 6-4 The relative impacts of the changes in temperature (A), solar radiation (B), and precipitation (C), and the overall impact of climate change (D) on winter wheat yield. 1–3 represents the period of short-term (2010–2039), medium-term (2040–2069), and long-term (2070–2099), respectively

Obviously, the negative impacts of warming temperatures could be counteracted by the positive impact of increasing radiation, and increasing precipitation contributes most to the total effects. Apart from CO_2 efficiency, the combined impact of warming temperatures, increasing radiation, and precipitation would increase wheat yield by 13.1%, 14.8%, and 29.1% (Table 6-5) over the short-term, medium-term, and long-term, respectively. Due to the negative impact of warming temperatures in sub-regions I to III, the increasing amplitude was lower than in other regions. Thus, the positive impact of climate change was higher in the southern part of the study region (Figure 6-4D).

Table 6-5 The relative impact of changes in temperature (F_T), solar radiation (F_S), and precipitation (F_P) in isolation, and the overall impact of climate change (F_{ALL}) on winter wheat yield for different sub-regions

Sub-region	F_T			F_S			F_P			F_{ALL}		
	S	M	L	S	M	L	S	M	L	S	M	L
I	−3.6	−8.2	−11.4	4.8	3.3	5.9	7.0	17.1	31.3	8.2	12.2	25.8
II	−10.4	−22.8	−30.2	7.0	8.4	14.7	7.5	17.8	23.6	4.1	3.4	8.0
III	−10.7	−22.1	−25.0	7.4	7.1	9.4	11.1	17.0	34.2	7.8	1.9	18.5
IV	−10.0	−16.9	−14.5	6.1	4.9	9.7	27.1	32.6	48.7	23.2	20.6	43.8
V	0.2	4.1	7.0	4.2	1.6	4.3	14.2	15.2	26.9	18.6	21.0	38.2
VI	7.0	17.8	24.8	1.7	3.8	2.4	5.7	3.6	7.5	14.5	25.3	34.9
Average	−3.6	−5.8	−5.8	5.0	4.3	7.1	11.7	16.3	27.7	13.1	14.8	29.1

6.3.5 Impact of elevated CO_2 on wheat yield

Compared to the normal CO_2 concentration of 380ppm, the increase in wheat yield tends to be greater with higher CO_2 concentration (Figure 6-5). The increasing amplitude was highest in sub-region V and lowest in sub-region II (Table 6-6).

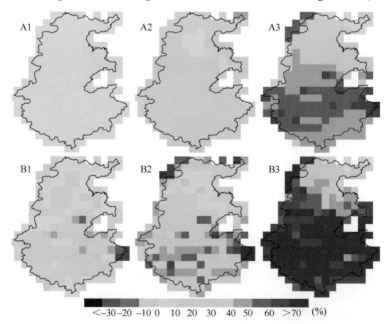

Figure 6-5 Relative impact of changes in concentration of carbon dioxide (A) and the impact of climate change with consideration of CO_2 fertilizer efficiency on winter wheat yield (B). 1–3 represents time period of short-term (2010–2039), medium-term (2040–2069), and long-term (2070–2099)

Table 6-6 Relative impacts of changes in concentration of carbon dioxide (F_C) and the overall impact of climate change on winter wheat yield with consideration of fertilizer efficiency of CO_2 (F_{ALL})

Sub-region	F_C			F_{ALL-C}		
	S	M	L	S	M	L
I	3.7	17.9	40.8	12.0	30.2	66.6
II	2.8	18.9	11.0	6.3	21.8	17.8
III	3.0	9.0	28.7	9.4	10.6	43.8
IV	2.4	8.8	30.6	16.8	21.6	57.8
V	6.4	30.3	90.1	37.7	65.6	154.3
VI	2.6	15.8	40.6	14.8	37.1	69.8
Average	3.6	17.4	41.8	16.8	32.3	70.9

Compared to the performance under the baseline concentration of CO_2, the average increases in amplitude are 3.6% (short-term, 423 ppm), 17.4% (medium-term, 571 ppm), and 41.8% (long-term, 798 ppm). When fertilizer efficiency based on CO_2 concentration is considered, overall climate change would elevate wheat yield by 16.8% (short-term), 32.3% (medium-term), and 70.9% (long-term).

6.4 Discussion

6.4.1 The impact of warming temperatures

Analyses of the climate-crop relationship have shown that global warming can have positive and negative effects, but negative impacts tend to dominate (Knox et al., 2012; Lobell and Field, 2007; Lobell et al., 2011; Schlenker and Lobell, 2010; Wang et al., 2008; You et al., 2009). The general consensus is that the impact of global warming has improved crop yields at high latitudes, but has deteriorated yields at low latitudes (Xiong et al., 2012). In this study, based on the CERES-Wheat model, warming temperatures in the Huang-Huai-Hai Plain under the future RCP8.5 scenario would reduce wheat yield by 3.6%, 5.8% and 5.8% for the short-term, medium-term and long-term, respectively, which is inconsistent with the previous consensus. By describing the empirical, statistical climate-crop relationship, Tao et al. (2014) demonstrated that warming climate in the past three decades increased wheat yield in Northern China by 0.9%–12.9%, but reduced wheat yield in Southern China by 1.2%–10.2%, with a large spatial difference. Using the CERES models, Xiong et al. (2012) also found that, due to the warming climate, cereal crops in China exhibited yield losses in areas at low latitudes, but had gains at high latitudes.

Obviously, increasing temperature during recent decades in the mid to high latitudes (such as the Huang-Huai-Hai Plain) has enriched thermal resources, but the cold stress events which threaten overwintering crops also decreased significantly (Tao et al., 2014; Xiao and Tao, 2014). Thus, the yield changes in the southern part of the Huang-Huai-Hai Plain showed an increasing trend (Figure 6-5A). On the other hand, increasing temperature would elevate crop water demands and drought risks by increasing potential evapotranspiration (Yang et al., 2015), which may counteract the advantageous increases in thermal resources. However, the results of many previous studies ignored the relationship between warming temperature and water status, by simulating crop yields under non-water stress conditions or building the statistical regression models based on yield data with full irrigation. In this study, yield was simulated under rain-fed conditions, and the effect of warming temperature on crop water demands was fully considered. If precipitation were kept unchanged, the water deficit condition (the ratio of precipitation minus evapotranspiration to evapotranspiration), which has already been the limiting factor in wheat production in the north part of Huang-Huai-Hai Plain, would deteriorate (Figure 6-6). However, in the southern part of the Huang-Huai-Hai Plain, due to sufficient precipitation, increasing evapotranspiration has an insignificant influence on the water deficit (Figure 6-6). Thus, in this study, the warming temperature with the precipitation and radiation remaining at the level of baseline period reduced wheat yield in the northern part of the Huang-Huai-Hai Plain, but increased it in the southern part of the Huang-Huai-Hai Plain.

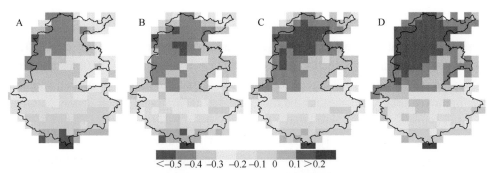

Figure 6-6 Water deficit condition (the ratio of precipitation minus evapotranspiration to evapotranspiration) of baseline (A), short-term(B), medium-term(C), and long-term(D) under increasing temperature with precipitation to be constant

6.4.2 Uncertainties

In this study, the CERES-Wheat model was used to estimate the impact of climate change on winter wheat yield in the Huang-Huai-Hai Plain. However, different crop models have different simulation algorithms to deal with leaf development, light interception, yield formation, crop phenology, and other factors (Palosuo et al., 2011). For example, in the CERES-Wheat model, the growth stage depends on thermal time, ignoring the stresses from water or fertilizer deficits. Also, CERES-Wheat only simulates nitrogen efficiency while the effects of phosphate and potassium were not considered. Thus, comparing different crop models can reveal and quantify the uncertainties related to crop growth and yield predictions (Liu et al., 2017; Tan et al., 2016; Yao et al., 2011). Comparisons of different crop models show that, in simulating yield responses to climate change, the variance between models would increase with warmer temperatures and higher concentration of CO_2 (Martre et al., 2015). Additionally, the fertilizer efficiency of CO_2 in the crop model was overestimated compared to that in the Free-Air CO_2 Enrichment (FACE) experiments (Tubiello et al., 2008). In this study, our results suggested that CO_2 enrichment would increase yield by 3.68%, 17.47%, and 41.87%, which was based on the mechanism from the Open-Top Chamber (OTC) or Environment Chamber. Thus, comparison analyses between crop models, especially the variance of CO_2 fertilizer efficiency, are necessary for quantification of the uncertainties related to crop models.

6.5 Conclusions

In this study, based on the CERES-Wheat model, changes in climate variables including temperature, solar radiation, and precipitation and the relative changes in the phenological phase of winter wheat during the growing season over the short-term (2010–2039), medium-term (2040–2069), and long-term (2070–2099) under the RCP8.5 scenario, compared to those observed over the baseline period (1981–2010), were analyzed. Further, the impact of the RCP8.5 climate scenario and CO_2 trends on winter wheat yield were untangled. Results showed that the shortened length of MDAP (planting to maturity) was equal to that of ADAP (planting to anthesis), indicating that changes in the length from anthesis to maturity were not apparent under the RCP8.5 scenario. Climate trend analysis showed that TMAX (maximum temperature), TMIN (minimum temperature), SRAD (solar radiation), and PRCP (precipitation) in the winter wheat growing season all increased. The trends of these climate variables would increase wheat yield by 13.1% (short-term), 14.8% (medium-term), and 29.1% (long-term). Increasing precipitation contributes most to the total climate change impact, elevating wheat yield by 11.7% (short-term), 16.3% (medium-term), and 27.7% (long-term), while the warming temperature decreased wheat yield by 3.6% (short-term), 5.8% (medium-term), and 5.8% (long-term), with the spatial distribution of a positive impact in the south and a negative impact in the north. Our analysis demonstrated that in the Huang-Huai-Hai Plain, which belongs to the mid-high latitude region, the advantageous increases in thermal resources would be counteracted with the disadvantageous water deficits caused by warming temperatures.

References

David B L, Christopher B F. 2007. Global scale climate-crop yield relationships and the impacts of recent warming. Environmental Research Letters, 2(1): 014002.

Hunt L A, Boote K J. 1998. Data for model operation, calibration, and evaluation. *In*: Tsuji G Y, Hoogenboom G, Thornton P K. Understanding Options for Agricultural Production. Dordrecht, Netherlands: Springer: 9-39.

Jones J W, Tsuji G Y, Hoogenboom G, Hunt L A, Thornton P K, Wilkens P W, Imamura D T, Bowen W T, Singh U. 1998. Decision support system for agrotechnology transfer: DSSAT v3. *In*: Tsuji G Y, Hoogenboom G, Thornton P K. Understanding Options for Agricultural Production. Dordrecht, Netherlands: Springer: 157-177.

Ju H, van der Velde M, Lin E, Xiong W, Li Y. 2013. The impacts of climate change on agricultural production systems in China. Climatic Change, 120(1-2): 313-324.

Keating B A, Carberry P S, Hammer G L, Probert M E, Robertson M J, Holzworth D, Huth N I, Hargreaves J N G, Meinke H, Hochman Z, McLean G, Verburg K, Snow V, Dimes J P, Silburn M, Wang E, Brown S, Bristow K L, Asseng S, Chapman S, McCown R L, Freebairn D M, Smith C J. 2003. An overview of APSIM, a model designed for farming systems simulation. European Journal of Agronomy, 18(3-4): 267-288.

Knox J, Hess T, Daccache A, Wheeler T. 2012. Climate change impacts on crop productivity in Africa and South Asia. Environmental Research Letters, 7: 034032.

Li Y, Huang H, Ju H, Lin E, Xiong W, Han X, Wang H, Peng Z, Wang Y, Xu J, Cao Y, Hu W. 2015. Assessing vulnerability and adaptive capacity to potential drought for winter-wheat under the RCP8.5 scenario in the Huang-Huai-Hai Plain. Agriculture, Ecosystems & Environment, 209: 125-131.

Liu X, Zhao Y, Shi X, Liu Y, Wang S, Yu D. 2017. Sensitivity and uncertainty analysis of CENTURY-modeled SOC dynamics in upland soils under different climate-soil-management conditions: A case study in China. Journal of Soils and Sediments, 17(1): 85-96.

Lobell D B, Field C B. 2007. Global scale climate–crop yield relationships and the impacts of resent warming. Enviromental Research Letters, 2(1): 014002.

Lobell D B, Schlenker W, Costa-Roberts J. 2011. Climate Trends and Global Crop Production Since 1980. Science, 333(6042): 616-620.

Martre P, Wallach D, Asseng S, Ewert F, Jones J W, Rotter R P, Boote K J, Ruane A C, Thorburn P J, Cammarano D, Hatfield J L, Rosenzweig C, Aggarwal P K, Angulo C, Basso B, Bertuzzi P, Biernath C, Brisson N, Challinor A J, Doltra J, Gayler S, Goldberg R, Grant R F, Heng L, Hooker J, Hunt L A, Ingwersen J, Izaurralde R C, Kersebaum K C, Mueller C, Kumar S N, Nendel C, O'Leary G, Olesen J E, Osborne T M, Palosuo T, Priesack E, Ripoche D, Semenov M A, Shcherbak I, Steduto P, Stoeckle C O, Stratonovitch P, Streck T, Supit I, Tao F, Travasso M, Waha K, White J W, Wolf J. 2015. Multimodel ensembles of wheat growth: Many models are better than one. Global Change Biology, 21(2): 911-925.

Mei X, Kang S, Yu Q, Huang Y, Zhong X, Gong D, Huo Z, Liu E. 2013. Pathways to synchronously improving crop productivity and field water use efficiency in the North China Plain. Scientia Agricultura Sinica, 46(6): 1149-1157.

Palosuo T, Kersebaum K C, Angulo C, Hlavinka P, Moriondo M, Olesen J E, Patil R H, Ruget F, Rumbaur C, Takac J, Trnka M, Bindi M, Caldag B, Ewert F, Ferrise R, Mirschel W, Saylan L, Siska B, Rotter R. 2011. Simulation of winter wheat yield and its variability in different climates of Europe: A comparison of eight crop growth models. European Journal of Agronomy, 35(3): 103-114.

Piao S, Ciais P, Huang Y, Shen Z, Peng S, Li J, Zhou L, Liu H, Ma Y, Ding Y, Friedlingstein P, Liu C, Tan K, Yu Y, Zhang T, Fang J. 2010. The impacts of climate change on water resources and agriculture in China. Nature, 467(7311): 43-51.

Ritchie J T, Singh U, Godwin D C, Bowen W T. 1998. Cereal growth, development and yield. *In*: Tsuji G Y, Hoogenboom G, Thornton P K. Understanding Options for Agricultural Production. Dordrecht, Netherlands: Springer: 79-98.

Schlenker W, Lobell D B. 2010. Robust negative impacts of climate change on African agriculture. Environmental Research Letters, 5(1): 014010.

Schlenker W, Roberts M J. 2009. Nonlinear temperature effects indicate severe damages to US crop yields under climate change. Proceedings of the National Academy of Sciences of the United States of America, 106(43): 15594-15598.

Shi W, Tao F, Zhang Z. 2013. A review on statistical models for identifying climate contributions to crop yields. Journal of Geographical Sciences, 23(3): 567-576.

Tan J, Cui Y, Luo Y. 2016. Global sensitivity analysis of outputs over rice-growth process in ORYZA model. Environmental Modelling & Software, 83: 36-46.

Tao F, Yokozawa M, Liu J, Zhang Z. 2008. Climate-crop yield relationships at provincial scales in China and the impacts of recent climate trends. Climate Research, 38(1): 83-94.

Tao F, Zhang Z, Xiao D, Zhang S, Rötter R P, Shi W, Liu Y, Wang M, Liu F, Zhang H. 2014. Responses of wheat growth and yield to climate change in different climate zones of China, 1981–2009. Agricultural and Forest Meteorology, 189-190(189): 91-104.

Timsina J, Humphreys E. 2006. Performance of CERES-Rice and CERES-Wheat models in rice-wheat systems: A review. Agricultural Systems, 90(1-3): 5-31.

Tubiello F, Schmidhuber J, Howden M, Neofotis P G, Park S, Fernandes E, Thapa D. 2008. Agriculture and rural development discussion Paper 42: Climate change response strategies for agriculture: Challenges and opportunities for the 21st century. The World Bank, Washington, DC: 63.

Wang H L, Gan Y T, Wang R Y, Niu J Y, Zhao H, Yang Q G, Li G C. 2008. Phenological trends in winter wheat and spring cotton in response to climate changes in northwest China. Agricultural and Forest Meteorology, 148(8-9): 1242-1251.

Williams J R, Renard K G, Dyke P T. 1983. EPIC: A new method for assessing erosion's effect on soil productivity. Journal of Soil and Water Conservation, 38(5): 381-383.

Wolfram S, David B L. 2010. Robust negative impacts of climate change on African agriculture. Environmental Research Letters, 5(1): 014010.

Xiao D, Moiwo J P, Tao F, Yang Y, Shen Y, Xu Q, Liu J, Zhang H, Liu F. 2015. Spatiotemporal variability of winter wheat phenology in response to weather and climate variability in China. Mitigation and Adaptation Strategies for Global Change, 20(7): 1191-1202.

Xiao D, Tao F. 2014. Contributions of cultivars, management and climate change to winter wheat yield in the North China Plain in the past three decades. European Journal of Agronomy, 52(Part

B): 112-122.

Xiong W, Holman I, Lin E D, Conway D, Li Y, Wu W B. 2012. Untangling relative contributions of recent climate and CO_2 trends to national cereal production in China. Environmental Research Letters, 7: 044014.

Xiong W, van der Velde M, Holman I P, Balkovic J, Lin E, Skalský R, Porter C, Jones J, Khabarov N, Obersteiner M. 2014. Can climate-smart agriculture reverse the recent slowing of rice yield growth in China? Agriculture, Ecosystems & Environment, 196: 125-136.

Yang J Y, Mei X R, Huo Z G, Zhao F H, Liu Q. 2015. Water consumption in summer maize and winter wheat cropping system based on SEBAL model in Huang-Huai-Hai Plain, China. Journal of Integrative Agriculture, 14(10): 2065-2076.

Yao F, Qin P, Zhang J, Lin E, Boken V. 2011. Uncertainties in assessing the effect of climate change on agriculture using model simulation and uncertainty processing methods. Chinese Science Bulletin, 56(8): 729-737.

Yong B, Ren L, Hong Y, Gourley J J, Chen X, Dong J, Wang W, Shen Y, Hardy J. 2013. Spatial-temporal changes of water resources in a typical semiarid basin of North China over the past 50 years and assessment of possible natural and socioeconomic causes. Journal of Hydrometeorology, 14: 1009-1034.

You L, Rosegrant M W, Wood S, Sun D. 2009. Impact of growing season temperature on wheat productivity in China. Agricultural and Forest Meteorology, 149(6-7): 1009-1014.

Yu Q, Li L, Luo Q, Eamus D, Xu S, Chen C, Wang E, Liu J, Nielsen D C. 2014. Year patterns of climate impact on wheat yields. International Journal of Climatology, 34(2): 518-528.

Yuan W, Cai W, Chen Y, Liu S, Dong W, Zhang H, Yu G, Chen Z, He H, Guo W, Liu D, Liu S, Xiang W, Xie Z, Zhao Z, Zhou G. 2016. Severe summer heatwave and drought strongly reduced carbon uptake in Southern China. Scientific Reports, 6(1): 18813.

Zhang X, Chen S, Liu M, Pei D, Sun H. 2005. Improved water use efficiency associated with cultivars and agronomic management in the North China Plain. Agronomy Journal, 97(3): 783-790.

Zhang X, Chen S, Sun H, Shao L, Wang Y. 2011. Changes in evapotranspiration over irrigated winter wheat and maize in North China Plain over three decades. Agricultural Water Management, 98(6): 1097-1104.

Chapter 7 Water consumption in winter wheat and summer maize cropping system based on SEBAL model in the Huang-Huai-Hai Plain, China

Abstract

Crop consumptive water use is recognized as a key element to understand regional water management performance. This study documents an attempt to apply a regional evapotranspiration model (SEBAL) and crop information to assessment of actual evapotranspiration (ETa) of a regional crop (winter wheat and summer maize) in the Huang-Huai-Hai Plain. The average seasonal ET_a of summer maize and winter wheat were 354.8 mm and 521.5 mm respectively in 3H Plain. A high ET_a belt of summer maize occurs in piedmont plain, while a low ET_a area was found in the hill-irrigable land and dry land area. For winter wheat, a high ET_a area was located in the middle part of 3H Plain, including low plain-hydropenia irrigable land and dry land, hill-irrigable land and dry land, and basin-irrigable land and dry land. Spatial analysis demonstrated a linear relationship between crop ET_a, normalized difference vegetation index (NDVI), and the land surface temperature (LST). A stronger relationship between ET_a and NDVI was found in the metaphase and last phase than other crop growing phase, as indicated by higher correlation coefficient values. Additionally, higher correlation coefficient was detected between ET_a and LST than that between ET_a and NDVI, and this significant relationship ran through the entire crop growing season. ET_a in the summer maize growing season showed a significant relationship with longitude, while ET_a in the winter wheat growing season showed a significant relationship with latitude. The results of this study will serve as baseline information for water resources management of 3H Plain.

7.1 Introduction

Agriculture is the largest water-consuming sector (FAO, 1994; Rosegrant et al., 2002) and irrigated agriculture has been expanding rapidly in many developing countries in recent decades, nearly doubling between 1962 and 1998 (Ali and Talukder, 2008; Carruthers et al., 1997). In the coming decades, water may become the most strategic resource, especially for agricultural production in arid and semi-arid regions of the world (Brewster et al., 2006), and water stress could threaten the sustainability of world agriculture. Accordingly, understanding the quantity of agricultural water consumption is a high priority in areas in which water is currently scarce and over-exploited (Perry, 2011). Evapotranspiration (ET) is a useful indicator of crop water consumption; therefore, accurate estimation of regional ET is essential for large scale water resources management (Rwasoka et al., 2011). Current estimates of actual evapotranspiration in China are mainly based on plot-scale experiments (Chen et al., 2002; Jiang and Zhang, 2004; Sun et al., 2003; Zhang et al., 1999), from the product of soil moisture and potential ET. However, such estimates are only useful for a specific area, and cannot be expanded to large-scale areas. The level of water consumption differs significantly across regions, farming systems, canal command areas, and farms (Molden et al., 2003). These differences come from many factors, including the source of irrigation water, farm management practices, the timing and efficiency of irrigation water in irrigated regions, and conservation tillage technologies, rainwater harvesting and cropping patterns in rainfed areas (Cai and Sharma, 2010).

Development of remote sensing technology has made it possible to estimate land surface evapotranspiration at the regional or basin scale. Numerous remote sensing methods for modeling actual evapotranspiration (ET_a) have been improved (De Oliveira et al., 2009; Jia et al., 2012; Teixeira and Bassoi, 2009; Teixeira et al., 2009). Bastiaanssen et al. (1998) were the first to use remote sensing data to estimate evapotranspiration in the Bhakra command area, India. And then, several studies demonstrated the strength of remote sensing in estimating crop evapotranspiration (Allen et al., 2007; Courault et al., 2005). Accurate estimation of agricultural water consumption is based on two inputs, the precise model, which has been calibrated and validated by many studies, and the ground truth information, including crop dominance maps, phenological characteristics, and agriculture productivity. However, ground truth information is often scarce and difficult to obtain.

The Huang-Huai-Hai Plain is the major crop producing region in China, with 3.5 million ha of highly intensive arable land, accounting for 19% of the country's crop production area. The recognized major limiting factor to crop production in the region is water shortage, which is expected to be exacerbated by increasing food demand in the region (Chen et al., 2005). Overexploitation of groundwater resources has resulted in water-table decrease at a rate of 1 m per year and severe groundwater depression in the past 20 years (Jia and Liu, 2002; Wang et al., 2009a). Moreover, climatic changes have intensified with an average decrease in rainfall of 2.92 mm per year(Liu et al., 2010). Thus, available agricultural water resources have become the most important factor influencing crop production in the Huang-Huai-Hai Plain, with the regional water scarcity situation becoming aggravated each year. Considering the spatial variation, accurate identification and region-wide water accounting are necessary for 3H Plain, for enabling reasonable allocation of the limitedly available agricultural water resources.

In recent years, Li et al. (2008) estimated the ET_a for winter wheat using the SEBAL (Surface Energy Balance Algorithm for Land) model and NOAA (National Oceanic and Atmospheric Administration) data for Hebei Province in the North China Plain (NCP). Yang et al. (2013b) analyzed the spatial and temporal variation of crop evapotranspiration (ET_c) and evapotranspiration of applied water (ET_{aw}) of summer maize during the growing season from 1960 to 2009 in the 3H farming region using the simulation of evapotranspiration of applied water (SIMETAW) model. However, observed phenological data were not considered in these investigations, and specific water consumption of winter wheat and summer maize has not yet been determined for larger areas, in the Huang-Huai-Hai Plain. In this research, an approach to estimate the actual evapotranspiration for winter wheat and summer maize respectively based on the MODIS data and SEBAL model is proposed and applied in 347 counties of the Huang-Huai-Hai Plain in China, which is a farming region providing about 61% and 31% of the nation's wheat and maize production, respectively (Ma et al., 2013; Wang et al., 2009a). The purpose of this study was: (1) to quantify actual evapotranspiration for winter wheat and summer maize, (2) to determine the spatial pattern of the ET_a of the two crops grown in the Huang-Huai-Hai Plain; and (3) to identify the relationship between crop ET_a and land surface parameters and geographic parameters. The findings from this research will provide useful information for agricultural water management practices in the Huang-Huai-Hai Plain, China.

7.2 Materials and methods

7.2.1 Study area

Huang-Huai-Hai Plain of Northern China is recognized as one of the largest plains in the country, extending from 31°14′ to 40°25′N and 112°33′ to 120°17′E, over an area of about 350,000 km². The climate is characterized by a temperate, sub-humid, and continental monsoon climate with an average annual precipitation of 500 mm to 800 mm (Ren et al., 2008). Winter is characterized by insufficient water for winter wheat development and production (Nguyen et al., 2011). Nevertheless, Huang-Huai-Hai

Plain is well accepted to be a major agricultural center, accounting for around 61% and 31% of China's wheat and maize production, respectively (Ma et al., 2013; Wang et al., 2009b). Accordingly, the cropping system in the plain is well-known to be a winter wheat-summer maize rotation system (Liang et al., 2011; Sun et al., 2011; Zhao et al., 2006). Currently, it is widely recognized that winter wheat is sown in early October and harvested in June of the second year, and that summer maize is then sown afterwards immediately and harvested in later September. Huang-Huai-Hai Plain is divided into six agricultural sub-regions, coastal land, a farming-fishing area (including the northern part, Zone 1, and the southern part, Zone 7), piedmont plain-irrigable land (Zone 2), low plain-hydropenia irrigable land and dry land (Zone 3), hill-irrigable land and dry land (Zone 4), basin-irrigable land and dry land (Zone 5) and hill-wet hot paddy-paddy field (Zone 6).

7.2.2 Crop dominance map

Ground truth missions were carried out in the Huang-Huai-Hai Plain in October 2011 and May 2012. The missions collected 175 samples from throughout the plain (not described in this book). Detailed crop patterns were recorded, including a crop mixture percentage visual estimate, crop growth period and past crop types (Cai and Sharma, 2010). The spectral signature curve of the summer maize-winter wheat rotation was extracted based on the sample points. ISODATA (Iterative Self-Organizing Data Analysis Technique) class identification technique and spectral matching technique (SMT) as proposed by Thenkabail et al. (2007b), were conducted to improve the summer maize-winter wheat rotation dominance map with ground truth information as the input. The cultivated area data of 347 counties were used for validation, and the R square value was 0.719, suggesting that the generated summer maize-winter wheat rotation dominance map was reliable.

7.2.3 Phenological data

The phenological data for the six agricultural sub-regions of the Huang-Huai-Hai Plain from 2011 to 2012 were acquired from the China Meteorological Administration (CMA). The data included the date of sowing and maturity of winter wheat and summer maize provided by the 50 agricultural meteorological stations in the Huang-Huai-Hai Plain. The average phenological date was calculated for the six agricultural sub-regions. Summer maize was sown from June 5 to June 20. Summer maize maturity was detected from the middle ten days to the last ten days of September. Winter wheat was sown during October, and harvested during the first ten days of June. Details regarding the phenological date of six agricultural zones are presented in Table 7-1.

Table 7-1 Average phenological data for winter wheat and summer maize in six sub-region area of Huang-Huai-Hai Plain

Agricultural zoning	Summer maize		Winter wheat	
	Sowing date	Maturity date	Sowing date	Maturity date
Coastal land-farming-fishing area(north)	6.15	9.25	10.1	6.15
Coastal land-farming-fishing area(south)	6.20	9.20	10.18	6.5
Piedmont plain-irrigable land	6.10	9.22	10.7	6.7
Low plain-hydropenia irrigable land and dry land	6.11	9.24	10.10	6.7
Hill-irrigable land and dry land	6.18	9.24	10.7	6.8
Basin-irrigable land and dry land	6.10	9.20	10.16	6.3
Hill-wet hot paddy-paddy field	6.5	9.16	10.27	5.25

7.2.4 MODIS products

MODIS products including MOD11A1 (land surface temperature/surface emissivity), MOD13A2 (NDVI) and MCD43B3 (surface albedo) were downloaded through NASA WIST for use in this study. The spatial resolution of the three MODIS products is 1 km. The temporal resolutions of MOD11A1, MOD13A2 and MCD43B3 were 1 d, 16 d, and 8 d, respectively. For land surface temperature images, cloudy areas were eliminated by replacing the values with the average of two images from the nearest clear dates (Cai and Sharma, 2010).

7.2.5 Meteorological data

Meteorological data are also needed for assessment of evapotranspiration. Datasets from 2011 to 2012 from 40 weather stations provided by the China Meteorological Administration (CMA) were used in this study (Table 7-1). The obtained data consisted of the daily observed maximum and minimum air temperature and wind speed measured at 10 m. Wind speed at 2 m was calculated from the wind speed at 10 m according to Allen et al. (2007) and interpolated with air temperature over the Huang-Huai-Hai Plain in pixels of 1000 m, which are needed for inputs in SEBAL model.

7.2.6 SEBAL model

SEBAL model introduction

In this research, the SEBAL model based on remote sensing technology was applied to estimate the daily ET. The MODIS data were used to estimate the regional ET for the study area. The calculation of the main parameters by the SABEL model is described below (Cai and Sharma, 2010).

The SEBAL model is based on the energy balance equation described by the following equation:

$$R_n = G + H + \lambda ET \qquad \text{Formula 7-1}$$

where, R_n is the net radiation(W·m^{-2}); G is the soil heat flux(W·m^{-2}); H is the sensible heat flux(W·m^{-2}); and λET is the latent heat flux associated with evapotranspiration (W·m^{-2}).

The net radiation flux on the land surface, R_n(W·m^{-2}), was calculated using the following equation:

$$R_n = (1-\alpha)K_{in} + (L_{in} - L_{out}) - (1-\varepsilon)L_{in} \qquad \text{Formula 7-2}$$

where, α is the surface albedo, K_{in} is the incoming short wave radiation (W·m^{-2}), L_{in} is the incoming long wave radiation (W·m^{-2}), L_{out} is the outgoing long wave radiation (W·m^{-2}), and ε is the land surface emissivity.

The soil heat flux is known to primarily depend on land surface characteristics and soil water content. The soil heat flux was calculated for the SEBAL model by the following equation:

$$G = \frac{T-273.16}{\alpha}\left[0.0032 \times \frac{\alpha}{0.9} + 0.0062 \times \left(\frac{\alpha}{0.9}\right)^2\right](1-0.98\text{NDVI}^4)R_n \qquad \text{Formula 7-3}$$

The sensible heat flux was calculated using the following equation:

$$H = \frac{\rho_{air} C_p dT}{r_{ah}} \qquad \text{Formula 7-4}$$

where, H is the sensible heat flux (W·m^{-2}), ρ_{air} is the air density (kg·m^{-3}), and C_P is the air specific heat at constant pressure (J·kg^{-1}·K^{-1}), r_{ah} is the aerodynamic impedance(S·m^{-1}).

Since the evaporative fraction Λ is constant during a day, the daily ET$_{24}$ (mm) can be estimated using the following equations:

$$\Lambda = \frac{\lambda ET}{R_n - G} \qquad \text{Formula 7-5}$$

$$ET_{24} = \frac{\Lambda(R_{24} - G_{24})}{\lambda} \qquad \text{Formula 7-6}$$

where, ET$_{24}$ is the daily net radiation (W·m^{-2}), R_{24} is the daily net radiation(W·m^{-2}), G_{24} is the daily soil heat flux (W·m^{-2}), and λ is the latent heat of vaporization (MJ·kg^{-1}). The SEBAL model is described in detail in Bastiaanssen et al. (1998).

Model validation

In this study, it was difficult to validate the ET$_a$ map because of its high variability and the low resolution produced by MODIS 1 km products. In recent years, the SEBAL model has been applied and validated in the Americas (Allen et al., 2002; Morse et al., 2000; Trezza, 2002), Europe (Jacob et al., 2002; Lagouarde et al., 2002), Africa (Bastiaanssen and Menenti, 1990; Farah and Bastiaanssen, 2001), and China (Li et al., 2008). Morse et al. (2000) reported that the error in daily ET was 15%, while that in monthly and quarterly ET estimation by SEBAL was 4% in Bear River Basin of Idaho. It was also reported that the error in the estimated daily ET was less than 7% in the Haihe basin (Xiong et al., 2006) and less than 8% in the middle region of Heihe basin (Wang et al., 2003). Taken together, these studies show that the SEBAL model has good efficiency and applicability for ET_a estimation. The model also works particularly well in the vegetative area including areas used for maize and wheat cropping, which was the focus of the present study (Cai and Sharma, 2010). Latent heat flux was extracted from the Yucheng station point, and Figure 7-1 shows the validation results with the field data for Yucheng station in Shandong Province. The correlation

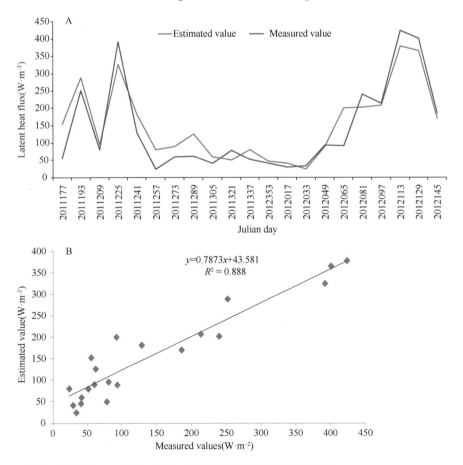

Figure 7-1 Validation results with the field data in Yucheng station
(A) Comparison between estimated and measured values; (B) Scatter map of estimated and measured values

coefficient between the estimated and measured values was 0.888, with a $P < 0.01$. Additionally, Table 7-2 compares the ranges of ET_a values from this study with those of previous studies. Taken together, these results show that SEBAL is suitable for estimating evapotranspiration in winter wheat and summer maize rotation in the Huang-Huai-Hai Plain.

Table 7-2 Comparison on ET_a of maize and wheat growing season for Huang-Huai-Hai Plain from this study and previous reports

Reference	Crop	ET_a (mm) This study	ET_a (mm) Literature	Reference
Yucheng, Shandong	wheat	456	(400–500) 450	Chen et al. (2012)
Yucheng, Shandong	maize	349	(300–370) 350	Chen et al. (2012)
Xinxiang, Henan	wheat	521	(374.9–551.7)	Xiao et al. (2009)
Piedmont plain	wheat	400–550	460	Ren and Luo (2004)
Piedmont plain	maize	300–500	390	Ren and Luo (2004)

7.3 Results

7.3.1 Crop ET_a

The ET_a of summer maize and winter wheat were calculated based on the crop dominant maps and phenological data. The ET_a map and histogram distribution, as well as its basic information are described in Figure 7-2 and Figure 7-3. The seasonal average ET_a of summer maize was 354.8 mm in the Huang-Huai-Hai Plain, with a minimum value of 239.4 mm and a maximum value of 552.3 mm. As shown in Figure 7-3-A, a high-ET belt occurs in the piedmont plain, from Beijing, Tianjin to the southern part of Hebei Province. The low ET_a area of summer maize was mainly found in the hill-irrigable land and dry land (Zone 4) area in Shandong Province. The total winter wheat ET_a was comparatively higher than the summer maize ET_a, with an average value of 521.5 mm. The maximum ET_a for winter wheat was 729.2 mm, which was found in the middle part of the Huang-Huai-Hai Plain, while the minimum value was 131.6 mm in the southeast part of Hebei Province. An ET_a belt between 500 mm and 600 mm was detected in more than 40% of the winter wheat cropping area, although significant variations within this range were observed. The difference from the summer maize ET_a map, higher ET_a area was mainly observed in the middle part of the Huang-Huai-Hai Plain, including low plain-hydropenia irrigable land and dry land (Zone 3), hill-irrigable land and dry land (Zone 4), and basin-irrigable land and dry land (Zone 5). Overall, the ET_a of summer maize and winter wheat displayed different spatial distributions among levels (Table 7-3).

Table 7-3 ET$_a$ of winter wheat and summer maize in study area

	Average ET$_a$ (mm)	Maximum ET$_a$ (mm)	Minimum ET$_a$ (mm)
Summer maize	354.8	552.3	239.4
Winter wheat	521.5	729.2	131.6

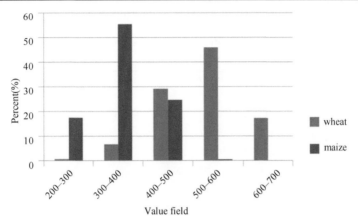

Figure 7-2 Value level distribution of ET$_a$ in winter wheat and summer maize season in the Huang-Huai-Hai Plain. The column chart was obtained based on the data at the size of 30″×30″ grid

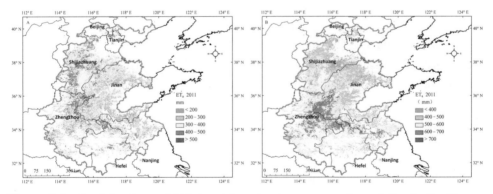

Figure 7-3 Spatial pattern of ET$_a$ in winter wheat-summer maize rotation in the Huang-Huai-Hai Plain, (A)spatial pattern of ET$_a$ in summer maize growing season; (B) spatial pattern of ET$_a$ in winter wheat growing season. These maps were obtained from the results based on SEBAL model

7.3.2 Correlation among ET$_a$, NDVI, and land surface temperature

Investigation of the relationships between crop ET$_a$ and the normalized difference vegetation index (NDVI) and the land surface temperature (LST) are helpful in understanding the effects of NDVI and LST changes on ET_a for winter wheat and summer maize in time series. Correlation coefficient analysis of the ET_a values of the two crops with NDVI and then LST were conducted for different Julian days. The ET_a value, NDVI and LST of the two crops were extracted from each raster in space. The relationships between ET_a and NDVI and ET_a and LST in the winter wheat and summer maize growing seasons are shown in Tables 7-4 and 7-5, respectively. NDVI is one of the most important parameters to the estimation of actual ET in many models.

Table 7-4 Relationship between ET_a in summer maize growing season and NDVI, and then LST through multivariate regression analysis

Julian day	Relationship between ET_a and NDVI							Relationship between ET_a and LST						
	1	2	3	4	5	6	7	1	2	3	4	5	6	7
2011193	0.22**	−0.04	0.11	−0.09	0.03	0.11	−0.05	−0.42**	−0.48**	−0.15*	−0.49**	−0.44**	−0.31**	−0.33**
2011209	0.32**	0.23**	0.24**	−0.11	0.00	−0.09	0.14	−0.31**	−0.25**	−0.10	−0.21**	−0.23**	−0.32**	−0.22**
2011225	0.14*	0.11	0.13	0.09	−0.08	0.19*	0.03	−0.28**	−0.48**	−0.18*	−0.45**	−0.31**	−0.41**	−0.34**
2011241	0.21**	0.30**	0.10	0.08	0.03	0.07	0.20**	−0.35**	−0.67**	−0.21**	−0.03	−0.19**	−0.17*	−0.28**
2011257	0.30**	0.26**	0.06	0.06	0.17*	0.21**	0.18*	−0.34**	−0.38**	0.16*	−0.47**	−0.28**	−0.08	−0.49**

Notes: * represents linear coefficients significant at $P < 0.05$. ** represents linear coefficients significant at $P < 0.01$

Table 7-5 Relationship between ET_a in winter wheat growing season and NDVI, and then LST through multivariate regression analysis

Julian day	Relationship between ET_a and NDVI							Relationship between ET_a and LST						
	1	2	3	4	5	6	7	1	2	3	4	5	6	7
2011289	0.01	0.12	−0.28**	0.02	−0.11	−0.14*	0.16*	−0.26**	−0.01	0.27**	−0.15*	0.06	−0.12	−0.26**
2011305	0.18*	0.17*	0.29**	0.12	−0.17*	−0.14*	−0.17*	−0.28**	−0.19**	−0.33**	−0.19**	−0.28**	−0.19**	−0.03
2011321	0.17*	0.06	0.35**	0.12	−0.04	−0.08	−0.01	−0.03	0.12	0.42**	0.14*	−0.45**	0.02	0.02
2011337	0.17*	0.11	0.43**	0.11	−0.08	−0.04	0.03	−0.38**	−0.25**	−0.70**	−0.30**	−0.36**	−0.07	0.05
2011353	0.19**	0.16*	0.44**	0.16*	−0.09	−0.01	0.05	−0.19**	0.20**	−0.24**	0.00	−0.42**	−0.16*	−0.45**
2012001	0.18*	0.15*	0.44**	0.16*	−0.07	−0.03	0.04	−0.24**	0.04	−0.36**	−0.16*	−0.45**	−0.40**	−0.30**
2012017	0.18	0.21	0.48	0.15	−0.12	0.06	0.07	−0.32**	−0.12	−0.22**	−0.17*	0.02	−0.19**	−0.10
2012033	0.15*	0.26**	0.56**	0.29**	−0.07	−0.06	0.19*	−0.45**	−0.02	0.09	−0.13	−0.29**	−0.27**	−0.05
2012049	0.18*	0.28**	0.54**	0.29**	−0.17*	−0.02	0.15*	−0.56**	−0.10	−0.15*	−0.48**	−0.15*	−0.22**	−0.15*
2012065	0.15	0.43**	0.55**	0.24**	−0.09	−0.02	0.32**	−0.52**	−0.38**	−0.35**	−0.45**	−0.13	−0.06	−0.29**
2012081	0.22**	0.36**	0.65**	0.19**	−0.05	0.02	0.09	−0.44**	−0.47**	−0.65**	−0.47**	−0.16*	−0.09	−0.26**
2012097	0.17	0.27**	0.67**	0.20**	0.07	−0.05	0.14*	−0.56**	−0.43**	−0.73**	−0.49**	0.05	−0.05	−0.36**
2012113	0.16	0.41**	0.66**	0.16	0.08	−0.03	0.11	−0.06	0.18*	−0.12	−0.32**	−0.27**	−0.18*	−0.28**
2012129	0.21	0.40**	0.57**	0.19**	0.14*	−0.13	0.12	−0.35**	−0.51**	−0.63**	−0.14	−0.30**	−0.47**	−0.50**
2012145	0.25**	0.31**	0.55**	0.14	0.32**	0.05	0.20**	−0.48**	−0.30**	−0.58**	−0.16*	−0.28**	−0.38**	−0.14*
2012161	0.23**	0.26**	0.03	0.21**	0.10	0.12	0.20**	−0.45**	−0.10	0.00	−0.14*	−0.07	−0.19**	0.06

Notes: * represents linear coefficients significant at $P < 0.05$. ** represents linear coefficients significant at $P < 0.01$

The results of this study showed that ET_a increased with NDVI. The linear relationship between ET_a and NDVI was consistent with the results of a previous study by Wang et al. (2012). NDVI in the metaphase and last phase was more close related to ET_a during the crop growing season, indicated by a higher positive correlation coefficient value. The positive relationship between ET_a and NDVI in the winter wheat growing season was closer than in the summer maize growing season in the metaphase and last phase, as indicated by a correlation coefficient value of less than 0.4 during the summer maize growing season but as high as 0.6 in the winter wheat growing season. This relationship appears to be closer in Zone 3 (piedmont plain-irrigable land), where the correlation coefficient values reached 0.67. The NDVI index increased significantly in the metaphase and last phase, when crop activities intensified. These changes were expressed as a lower impedance of evapotranspiration and increased latent heat flux in each pixel. This physiological reaction from the crop may have led to an increase in the actual ET. These changes were more obvious in flat interior regions, such as piedmont plain-irrigable land, than in basins or hills. The relationship between ET_a and NDVI was closely during winter and spring, because there was less precipitation participating in the space hydrological cycle.

As described above, a higher value of correlation coefficient was detected between ET_a and LST than between ET_a and NDVI. The significant relationship between ET_a and LST ran through the entire crop growing season. Additionally, the correlation between ET_a and LST was negative, indicating that an increased LST may lead to a decreased ET_a. In the later portion of the winter wheat growing season in Zone 3 (piedmont plain-irrigable land), the correlation coefficient value became higher ($R>0.7$). Temperature is an important factor which is associated with stomatal conductance and transpiration (Yang et al., 2012). For maize, the effect of growth temperature on transpiration was obvious when maize was grown at low temperature (22/18°C) and measured at higher temperature (30°C). The Huang-Huai-Hai Plain is acknowledged as a water-stress area that primarily receives rainfall during summer. As a results, serious drought always occurs in winter and spring, and may impact transpiration and canopy temperature. When crops are subjected to water stress, they close their leaf stomata, which reduces evapotranspiration, leading to increased crop canopy temperature.

7.3.3 Correlation between ET_a and geographic parameters

For a given region, reference evapotranspiration (ET_0) is only determined by weather parameters; however, several factors can affect actual evapotranspiration, such as soil types, current precipitation, crop types, soil water storage in the early stage, and field management. We attempted to identify the relationship between ET_a and geographic parameters in this study because they may reflect climate fluctuations, changes on soil characteristics, field management and irrigation schemes with geographic transitions. ET_a in the summer maize period represented a significant relationship with longitude ($P<0.01$), which described an increasing trend from the eastern portion to the western part of the Huang-Huai-Hai Plain (Figure 7-4-A). This spatial pattern of ET_a in the summer maize growing season is in accordance with that of precipitation. Usually, the growing season for summer maize in the Huang-Huai-Hai Plain is from July to September, when there is concentrated precipitation and higher temperature. During this period, less than 20% of ET_a is from irrigation. When compared to temperature, crop physiology is more sensitive to water for summer maize owing to the sufficient heat resources in the summer maize growing season. Rainfall was considered as the

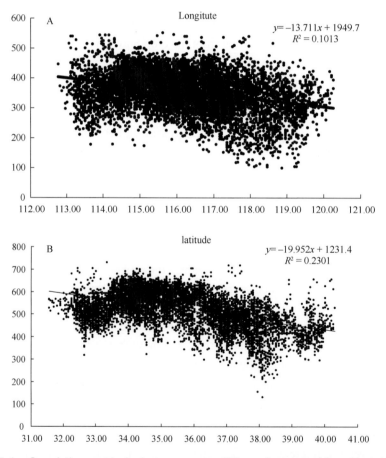

Figure 7-4 Correlation analysis between crop ET_a and geographic parameters. A is correlation analysis between ET_a in summer maize growing season and longitude. X-axis represents the longitude and Y-axis represents ET_a (mm) for summer maize growing season. B is correlation analysis between ET_a in winter wheat growing season and latitude. X-axis represents the latitude and Y-axis represents ET_a (mm) for winter wheat growing season

main crop water resource in the eastern portion of the Huang-Huai-Hai Plain, where more precipitation was detected in the past 40 years. Supplementary irrigation has always been used in the western part of the region, which is characterized by piedmont plain-irrigable land, low plain-hydropenia irrigable land and dry land.

As shown in Figure 7-4-A, the ET_a for winter wheat had a significant relationship with latitude (with $R^2=0.23$, $P<0.01$), increasing as latitude increased. Winter wheat accounts for about 70% of the total agricultural water use in this area, and precipitation during the winter wheat growing season ranges from 100 mm to 180 mm (Li et al., 2010), which can only meet around 25%–40% of the water requirements for the season. Although it is an important area for winter wheat production, rainfall in the region is erratic and limited during the growing stage; accordingly, supplementary irrigation has been widely adopted to ensure maximum production (Li et al., 2008; Sun et al., 2006). For irrigated wheat, seasonal ET_a mostly ranges from 400 mm to 600 mm, 70% of which is derived from irrigation. Spatial differences have not only been found in precipitation, but also in irrigation practices. According to Yang et al. (2013a), precipitation in the southeastern part of the Huang-Huai-Hai Plain can satisfy over 50% of the total water consumption for winter wheat. However, irrigation was identified as the dominant water resource in the northern part of the Huang-Huai-Hai Plain, including piedmont plain-irrigable land (Zone 2) and low plain-hydropenia irrigable land and dry land (Zone 3), where irrigation accounts for more than 60% of the winter wheat water consumption.

7.4 Discussion

7.4.1 Assessment of regional crop evapotranspiration

The findings presented in this chapter are the first region-wide, integrated remote sensing, SEBAL model, and ground truth and phenological data to estimate ET_a in the Huang-Huai-Hai Plain. There are various ways to estimate crop water consumption, most of which are relatively precise at very small scales, but impractical over large scales. It is necessary to identify crop water consumption as the main consequence in agricultural hydrological processes, not only at experimental field points in a controlled environment, but also at regional scales. At regional scales, crop evapotranspiration is often more relevant to policy, agricultural input, soil types and available resources. However, water consumption by crops cannot be accurately identified without crop distribution information (Cai and Sharma, 2010). This study is the first attempt to apply regional evapotranspiration model and crop information to assessment of crop (winter wheat and summer maize cropping system) evapotranspiration at a large regional scale, as in the Huang-Huai-Hai Plain. This addition marks the improvement of this research work over numerous previous studies (Cai and Sharma, 2010; Immerzeel et al., 2008; Zwart and Bastiaanssen, 2007; Zwart et al., 2010).

7.4.2 Separation of evapotranspiration of the two crops

A method was developed to extend daily evapotranspiration to the crop growing season in this study. The extrapolation of daily evapotranspiration to crop growing seasons in pixel level was conducted through spatial interpolation methods based on crop phenological data. Crop (winter wheat and summer maize) phenological data for the Huang-Huai-Hai Plain, as well as other essential information, was used to obtain pixel crop phenological data. Crop growth is more closely related to latitude and longitude, elevation, crop varieties and meteorological factors, such as air temperature, light and water (Yang et al., 2011). However, the method explained in this paper avoided complex factors and parameters above, and can be easily applied elsewhere. Additionally, evapotranspiration extracted from winter wheat and summer maize can be more accurately estimated than the factors above.

7.4.3 Possible uncertainty of results

The ET_a in this study was 354.8 mm and 521.5 mm for summer maize and winter wheat respectively, which is higher than those in previous studies (Chen et al., 2012; Xiao et al., 2009). It is not surprising that ET_a was lower at the research stations since they are operated under a controlled environment to achieve the maximum water use efficiency, and are less constrained than farmers with regards to resources availability (Yan and Wu, 2014). In general, the ET_a of winter wheat was higher than that of summer maize in the Huang-Huai-Hai Plain. These findings partially agreed with those of Ren and Luo (2004) and Chen et al. (2012) who pointed out that physiological characteristics of crops, field management measures and irrigation programs are major factors influencing ET_a, even though reference evapotranspiration (ET_0) and crop water requirement (ET_c) occasionally show different characteristics. However, spatial differences have made possible contrary results in certain region. For example, in Hebei Province the ET_a of summer maize was slightly higher than that during the winter wheat growing season. Spatial analysis demonstrated a linear relationship between crop ET_a and NDVI/land surface temperature, which is consistent with the result of a study conducted by Wang et al. (2012). It should be noted that the raster pixel was upscaled to 1000 m×1000 m for easier presentation, which may have caused the pixels to merge together, and decreased the relevance between dependent and independent variables. As a result, the correlation coefficient between crop ET_a/NDVI and land surface temperature may actually be higher than the calculated value.

7.4.4 Need for refinement

It is important to note that there are some uncertainties associated with estimating crop evapotranspiration using remote sensing data and the SEBAL model over a large-scale region such as the Huang-Huai-Hai Plain. Gathering remote sensing data is a complicated process that must be followed by sensor calibration and atmospheric correction (Cai and Sharma, 2010). The spatial distribution of evapotranspiration modeling by SEBAL is 1000 m×1000 m; however, mixed cropping patterns and fragmented farming are found common in crop planting extraction research, so sub-pixel area fraction estimation is well accepted (Hao et al., 2011; Thenkabail et al., 2007a). In some situations, one pixel contains several but excludes target crop, such as water body, residential areas, and forest land. Under the given conditions, the image element may be exaggerated or ignored, which can lead to poor estimations and increased errors.

7.5 Conclusions

In this study, actual evapotranspiration for winter wheat and summer maize respectively and its spatial patterns were quantified in the Huang-Huai-Hai Plain. The seasonal average ET_a of summer maize and winter wheat were 354.8 mm and 521.5 mm in the Huang-Huai-Hai Plain. A high-ET belt of summer maize covers the piedmont plain, and low ET_a areas of summer maize are mainly found in the hill-irrigable land and dry land area. For winter wheat, higher ET_a areas were located in the middle part of the Huang-Huai-Hai Plain, including low plain-hydropenia irrigable land and dry land (Zone 3), hill-irrigable land and dry land (Zone 4), and basin-irrigable land and dry land (Zone 5). Spatial analysis demonstrated a linear relationship between crop ET_a and NDVI, as well as between ET_a and land surface temperature. During the crop growing season, ET_a was more closely related to NDVI in the metaphase and last phase.

We attempted to identify relationships between ET_a and land surface parameters and geographic parameters. NDVI in the metaphase and last phase showed a closer correlation to ET_a in the crop growing season, and a significant relationship between ET_a and LST was observed throughout the crop growing season. ET_a in the summer maize growing season was correlated with longitude, while ET_a in the winter wheat growing season showed a significant relationship with latitude. Field management (supplemental irrigation) also showed a strong response to the ET_a pattern in the Huang-Huai-Hai Plain.

References

Ali M H, Talukder M S U. 2008. Increasing water productivity in crop production–a synthesis. Agricultural Water Management, 95(11): 1201-1213.

Allen R G, Morse A, Tasumi M, Trezza R, Bastiaanssen W, Wright J L, Kramber W. 2002. Evapotranspiration from a satellite-based surface energy balance for the Snake Plain Aquifer in Idaho. In: Proceedings of the USCID/EWRI Conference on Energy, Climate, Environment and Water Issues and opportunities for Irrigation and Drainage. San Luis Obispo, California, USA: 167-178.

Allen R G, Tasumi M, Trezza R. 2007. Satellite-based energy balance for mapping evapotranspiration with internalized calibration (METRIC)–Model. Journal of Irrigation and Drainage Engineering, 133(4): 380-394.

Bastiaanssen W, Menenti M, Feddes R A, Holtslag A. 1998. A remote sensing surface energy balance algorithm for land (SEBAL). 1. Formulation. Journal of Hydrology, 212(1-4): 198-212.

Bastiaanssen W, Menenti M. 1990. Mapping groundwater losses in the western desert of Egypt with satellite measurements of surface reflectance and surface temperature. Proceedings and Information—TNO Committee on Hydrological Research, 42: 61-90.

Brewster M, Herrmann T, Bleish B, Pearl R. 2006. A Gender Perspective on Water Resources and Sanitation. Wagadu, 3: 1-23.

Cai X L, Sharma B R. 2010. Integrating remote sensing, census and weather data for an assessment of rice yield, water consumption and water productivity in the Indo-Gangetic river basin. Agricultural Water Management, 97(2): 309-316.

Carruthers I, Rosegrant M W, Seckler D. 1997. Irrigation and food security in the 21st century. Irrigation and Drainage Systems, 11(2): 83-101.

Chen B, Ou Y, Chen W, Liu L. 2012. Water consumption for winter wheat and summer maize in the North China plain in recent 50 years. Journal of Natural Resources, 27: 1186-1199. (in Chinese)

Chen J, Tang C, Sakura Y, Yu J, Fukushima Y. 2005. Nitrate pollution from agriculture in different hydrogeological zones of the regional groundwater flow system in the North China Plain. Hydrogeology Journal, 13(3): 481-492.

Chen S, Zhang X, Liu M. 2002. Soil temperature and soil water dynamics in wheat field mulched with maize straw. Chinese Journal of Agrometeorology, 23(4): 34-37. (in Chinese)

Courault D, Seguin B, Olioso A. 2005. Review on estimation of evapotranspiration from remote sensing data: From empirical to numerical modeling approaches. Irrigation and Drainage systems, 19(3-4): 223-249.

De Oliveira A S, Trezza R, Holzapfel E A, Lorite I, Paz V P S. 2009. Irrigation water management in Latin America. Chilean Journal of Agricultural Research, 69: 7-16.

FAO (Food and Agriculture Organization). 1994. The state of food and agriculture 1993. Food and Agriculture Organization. Rome, Italy.

Farah H O, Bastiaanssen W. 2001. Spatial variations of surface parameters and related evaporation in the Lake Naivasha Basin estimated from remote sensing measurements. Hydrological Processes, 15: 1585-1607.

Hao W P, Mei X R, Cai X L, Du J T, Liu Q. 2011. Crop planting extraction based on multi-temporal remote sensing data in Northeast China. Transactions of the CSAE, 27(1): 201-207. (in Chinese)

Immerzeel W W, Gaur A, Zwart S J. 2008. Integrating remote sensing and a process-based hydrological model to evaluate water use and productivity in a south Indian catchment. Agricultural Water Management, 95(1): 11-24.

Jacob F, Olioso A, Gu X F, Su Z, Seguin B. 2002. Mapping surface fluxes using airborne visible, near infrared, thermal infrared remote sensing data and a spatialized surface energy balance model. Agronomie, 22(6): 669-680.

Jia J, Liu C. 2002. Groundwater dynamic drift and response to different exploitation in the North China Plain: A case study of Luancheng County, Hebei Province. Acta Geographica Sinica-Chinese Edition, 57(2): 201-209.

Jia Z, Liu S, Xu Z, Chen Y, Zhu M. 2012. Validation of remotely sensed evapotranspiration over the Hai River Basin, China. Journal of Geophysical Research Atmospheres, 17(D13).

Jiang J, Zhang Y. 2004. Soil-water balance and water use efficiency on irrigated farmland in the North China Plain. Journal of Soil and Water Conservation, 18(3): 61-65. (in Chinese)

Lagouarde J, Jacob F, Gu X F, Olioso A, Bonnefond J, Kerr Y, Mcaneney K J, Irvine M. 2002. Spatialization of sensible heat flux over a heterogeneous landscape. Agronomie, 22(6): 627-634.

Li H, Zheng L, Lei Y, Li C, Liu Z, Zhang S. 2008. Estimation of water consumption and crop water productivity of winter wheat in North China Plain using remote sensing technology. Agricultural Water Management, 95(11): 1271-1278.

Li Q, Dong B, Qiao Y, Liu M, Zhang J. 2010. Root growth, available soil water, and water-use efficiency of winter wheat under different irrigation regimes applied at different growth stages in North China. Agricultural Water Management, 97(10): 1676-1682.

Liang W, Carberry P, Wang G, L R, L H, Xia A. 2011. Quantifying the yield gap in wheat-maize cropping systems of the Hebei Plain, China. Field Crop Research, 124(2): 180-185.

Liu S, Mo X, Lin Z, Xu Y, Ji J, Wen G, Richey J. 2010. Crop yield responses to climate change in the Huang-Huai-Hai Plain of China. Agricultural Water Management, 97(8): 1195-1209.

Ma Y, Feng S Y, Song X F. 2013. A root zone model for estimating soil water balance and crop yield response to deficit irrigation in the Northern China Plain. Agricultural Water Management, 127: 13-24.

Molden D, Murray-Rust H, Sakthivadivel R, Makin I. 2003. A water-productivity framework for understanding and action. *In*: Tuong T P, Bouman B A M. Water Productivity in Agriculture: Limits and Opportunities for Improvement. Wallingford, UK: CABI Publishing: 1-18.

Morse A, Tasumi M, Allen R G, Kramber W J. 2000. Application of The SEBAL Methodology for Estimating Consumptive Use of Water and Streamflow Depletion in the Bear River Basin of Idaho Through Remote Sensing. Idaho Department of Water Resources, University of Idaho, UAS.

Nguyen T T, Qiu J J, Ann V, Li H, Eric V R. 2011. Temperature and precipitation suitability evaluation for winter wheat and summer maize cropping system in the Huang-Huai-Hai plain of China. Journal of Integrative Agriculture, 10(2): 275-288.

Perry C. 2011. Accounting for water use: Terminology and implications for saving water and increasing production. Agricultural Water Management, 98(12): 1840-1846.

Ren H, Luo Y. 2004. The experimental research on the water-consumption of winter wheat and summer maize in the Northwest plain of Shandong province. Journal of Irrigation and Drainage, 23: 37-39. (in Chinese)

Ren J, Chen Z, Zhou Q, Tang H. 2008. Regional yield estimation for winter wheat with MODIS-NDVI data in Shandong, China. International Journal of Applied Earth Observation and Geoinformation, 10(4): 403-413.

Rosegrant M W, Cai X M, Cline S A. 2002. Global water outlook to 2025: Averting an impending crisis. A 2020 Vision for food, agriculture, and environment initiative. International Food Policy Research Institute, Washington, D.C., USA.

Rwasoka D T, Gumindoga W, Gwenzi J. 2011. Estimation of actual evapotranspiration using the Surface Energy Balance System (SEBS) algorithm in the Upper Manyame catchment in Zimbabwe. Physics and Chemistry of the Earth (Parts A/B/C), 36(14-15): 736-746.

Sun H, Liu C, Zhang X, Shen Y, Zhang Y. 2006. Effects of irrigation on water balance, yield and WUE of winter wheat in the North China Plain. Agricultural Water Management, 85(1-2): 211-218.

Sun H, Zhang Y, Zhang X, Mao X, Pei D, Gao L. 2003. Effects of water stress on growth and development of winter wheat in the North China Plain. Acta Agriculture Boreali-Sinica, 18: 23-26. (in Chinese)

Sun Q, Krbel R, Mller T, Rmheld V, Cui Z, Zhang F, Chen X. 2011. Optimization of yield and water-use of different cropping systems for sustainable groundwater use in North China Plain. Agricultural Water Management, 98(5): 808-814.

Teixeira A D C, Bassoi L H. 2009. Crop water productivity in semi-arid regions: From field to large scales. Annals of Arid Zone, 48(3): 1-13.

Teixeira A H de C, Bastiaanssen W G M, Ahmad M D, Bos M G. 2009. Reviewing SEBAL input parameters for assessing evapotranspiration and water productivity for the Low-Middle São Francisco River basin, Brazil: Part A: Calibration and validation. Agricultural and Forest Meteorology, 149(3): 477-490.

Thenkabail P S, Biradar C M, Noojipady P, Cai X L, Dheeravath V, Li Y J, Velpuri M, Gumma M K, Pandey S. 2007a. Sub-pixel area calculation methods for estimating irrigated areas. Sensors, 7(11): 2519-2538.

Thenkabail P S, GangadharaRao P, Biggs T, Krishna M, Turral H. 2007b. Spectral matching techniques to determine historical land-use/land-cover (LULC) and irrigated areas using time-series 0.1-degree AVHRR Pathfinder datasets. Photogrammetric Engineering & Remote Sensing, 73: 1029-1040.

Trezza R. 2002. Evapotranspiration using a satellite-based surface energy balance with standardized ground control. Ph D thesis, Utah State University, Logan, UT.

Wang E, Chen C, Yu Q. 2009b. Modeling the response of wheat and maize productivity to climate variability and irrigation in the North China Plain. In: 18th World IM4CSIMODSIM Congress. Cairns, Australia: 2742-2748.

Wang J, Gao F, Liu S. 2003. Remote sensing retrieval of evapotranspiration over the scale of drainage basin. Remote Sensing Technology and Application, 18(5): 332-338. (in Chinese)

Wang S, Song X, Wang Q, Xiao G, Liu C, Liu J. 2009a. Shallow groundwater dynamics in North China Plain. Journal of Geographical Sciences, 19(2): 175-188.

Wang W T, Zhao Q L, Du J. 2012. Advances in the study of evapotranspiration of regional land surface based on remote sensing technology. Remote Sensing for Land and Resources, 24(1): 1-7. (in Chinese)

Xiao J, Liu Z, Duan A. 2009. Studies on water production function of winter wheat in Xinxiang district. Journal of Henan Agricultural, 1: 55-59. (in Chinese)

Xiong J, Wu B, Zhou Y, Li J. 2006. Estimating evapotranspiration using remote sensing in the Haihe basin. International Geoscience and Remote Sensing Symposium, 1044-1047.

Yan N, Wu B. 2014. Integrated spatial-temporal analysis of crop water productivity of winter wheat in Hai Basin. Agricultural Water Management, 133: 24-33.

Yang J, Liu Q, Mei X, Yan C, Ju H, Xu J. 2013a. Spatiotemporal characteristics of reference evapotranspiration and its sensitivity coefficients to climate factors in Huang-Huai-Hai plain, China. Journal of Integrative Agriculture, 12(12): 2280-2291.

Yang J, Mei X, Liu Q, Yan C, He W, Liu E, Liu S. 2011. Variations of winter wheat growth stages under climate changes in Northern China. Chinese Journal of Plant Ecology, 35(6): 623-631. (in Chinese)

Yang X L, Gao W S, Shi Q H, Chen F, Chu Q Q. 2013b. Impact of climate change on the water requirement of summer maize in the Huang-Huai-Hai farming region. Agricultural Water Management, 124: 20-27.

Yang Z, Sinclair T R, Zhu M, Messina C D, Cooper M, Hammer G L. 2012. Temperature effect on transpiration response of maize plants to vapour pressure deficit. Environmental and Experimental Botany, 78: 157-162.

Zhang H, Wang X, You M, Liu C. 1999. Water-yield relations and water-use efficiency of winter wheat in the North China Plain. Irrigation Science, 19(1): 37-45.

Zhao R F, Chen X P, Zhang F S, Zhang H, Schroder J, Rmheld V. 2006. Fertilization and nitrogen balance in a wheat-maize rotation system in North China. Agronomy Journal, 98(4): 938-945.

Zwart S J, Bastiaanssen W G, de Fraiture C, Molden D J. 2010. WATPRO: A remote sensing based model for mapping water productivity of wheat. Agricultural Water Management, 97(10): 1628-1636.

Zwart S J, Bastiaanssen W G. 2007. SEBAL for detecting spatial variation of water productivity and scope for improvement in eight irrigated wheat systems. Agricultural Water Management, 89(3): 287-296.

Chapter 8 An assessment of water consumption, grain yield and water productivity of winter wheat in agricultural sub-regions of Huang-Huai-Hai Plain, China

Abstract

The Huang-Huai-Hai Plain, the wheat production base of China, is recognized to be one of the most water stressed areas of the world due to the excessive exploitation of groundwater combined with precipitation reduction. Identification of areas with a low crop water productivity (CWP) and achieving a higher grain yield per unit of consumed water are therefore of uttermost importance for the future agriculture in this region. In this paper, the crop water productivity of winter wheat (*Triticum aestivum* L.) was estimated and subsequently analyzed by combining remote sensing imagery and county-level census and meteorological data of six agricultural sub-regions of the Huang-Huai-Hai Plain. The average CWP of winter wheat in the plain is 0.95 kg·m^{-3}, with a range of 0.24 kg·m^{-3} to 1.99 kg·m^{-3} for three periods, namely 2001, 2006, and 2011. The spatial analysis of the relationship among CWP, grain yield, and actual evapotranspiration (ET$_a$) describes a close linear relationship between CWP, grain yield, and ET$_a$ across the Huang-Huai-Hai Plain. The grain yield increases to a critical value of 520 mm and 480 mm for ET$_a$ in the northern and southern zones, respectively, and then drops as the ET$_a$ increases. The temporal analysis indicates an increase of the CWP and yield by an average of 0.25 kg·m^{-3} and 0.09 t·ha^{-1}, respectively, and an average reduction of 82.1 mm for ET$_a$ during the studied periods. It is concluded that the improvements of the CWP are due to the grain yield increment as a result of better cultivars, fertilizer improvement and other management practices rather than water consumption reductions. The results are expected to provide basic information for the agricultural water management, improvement of CWP, and choice of adaptive mechanisms in water-scarce regions.

8.1 Introduction

The worldwide water demand has increased substantially over the past decades due to a rapid economic development, population growth, and increasing water scarcity in many parts of the world (Vrsmarty et al., 2000). The growing water scarcity in agriculture at the global scale has become a bottleneck of the ecological, environmental, and socioeconomic development and sustainability of agriculture (Immerzeel et al., 2010; Murray et al., 2012; Wolf et al., 2003; Yoffe et al., 2003). Agriculture is identified as the largest water-consuming sector and in consequence, irrigated agriculture has expanded rapidly over recent decades in many developing countries (Ali and Talukder, 2008; Carruthers et al., 1997). Irrigated agriculture contributes between around 25% and 50% to global food production (Boutraa, 2010). An FAO analysis of 93 developing countries indicated that agricultural production would be expected to increase over the period of 1998–2030 by 49% in rainfed systems and by 81% in irrigated systems (Playán and Mateos, 2006). To achieve a sustainable and higher grain yield, irrigation is indispensable, making the water shortage problem more serious (Li et al., 2012). Understanding how the productivity of water can be increased is a high priority in regions with scarce and/or over–exploited water resources (Perry, 2011).

The crop water productivity (CWP) is defined by the amount of produced output per unit of water consumption under a specific condition (Cai et al., 2011; Molden et al., 2007, 2010), and becomes a critical indicator for the quantification of the impact of irrigation scheduling concerning water management (Igbadun et al., 2006). Quantitative information on the CWP estimated from grain yield and water consumption is consequently essential for the irrigation water management strategies in a particular region. A number of researchers reported the CWP in China, with varied spatial and temporal resolutions based on model simulation, estimation from remote sensing data and calculation from collected irrigation district data (Cao et al., 2015; Huang and Li, 2010; Liu et al., 2007; Rosegrant et al., 2002; Yan and Wu, 2014).

Most of the oforementioned studies reported the relationship between water consumption, grain yield, and CWP, and suggests that it is important to consider general situation in the controlled environment and related decisions (Zwart and Bastiaanssen, 2004). Zhang et al. (2011) reported that winter wheat biomass has a closer linear relationship with actual evapotranspiration (ET_a) than summer maize based on experimental research from 1979–2009. Perry et al. (2009) reported that the relationship between biomass and transpiration is essentially linear for given crops and climate conditions based on a review. However, other results suggest that the relationship between ET and yield is not always linear. Li et al. (2012) indicated the relationship between grain yields and seasonal ET was best described by a quadratic function obtained by regression analysis in Southwest China. Li et al. (2009) reported a relationship between ET and yield in the form of a parabola; Which, however was not statistically significant.

The Huang-Huai-Hai Plain is known as the largest agricultural production region in China, where the most important factor limiting agricultural production is water. This problem will become more severe in the future due to the increasing food demand, soil quality deterioration, and diminishing water quality. Hence, the Huang-Huai-Hai Plain faces a serious threat of water scarcity due to excessive exploitation of groundwater combined with a precipitation reduction (Jia and Liu, 2002; Liu et al., 2010b). The key point to maintain high crop productivity and reduce the depletion of water resources is to improve the CWP in the Huang-Huai-Hai Plain. Nevertheless, the relationship between the water consumption and grain yield of the agricultural sub-regions in the Huang-Huai-Hai Plain and the critical ET_a value at which the grain yield begins to drop remain unknown up to now.

Therefore, this chapter focuses on (1) a quantification of field scale productivity of wheat using a satellite-based model, linear regression equation-integrated remote sensing data, and census data; (2) a spatially differentiated characterisation of the water productivity of wheat; and (3) a discussion of potential improvements of the water productivity for wheat using the structure of water consumption and an analysis of agronomical practice across the Huang-Huai-Hai Plain.

8.2 Materials and methods

8.2.1 Study region description

The Huang-Huai-Hai Plain is generally acknowledged as one of the largest plains in China, located in the temperate, sub-humid, and continental monsoon climate zone with a cumulative temperature (>0°C) of 4200°C to 5500°C and average annual precipitation range of 500 mm to 800 mm (Ren et al., 2008). Although the precipitation is insufficient for cultivation, the Huang-Huai-Hai Plain remains one of the main crop production centres in China, providing 31% and 61% of the nation's maize and wheat production respectively (Wang et al., 2009). Typically, the main crop system is characterized by maize-wheat rotation with two harvests per year (Liang et al., 2011; Sun et al., 2011; Zhao et al., 2006).

8.2.2 Data collection

The China Meteorological Administration (CMA) provided us with a historical dataset acquired between 2001 and 2015, which was composed of climatic variables from 40 meteorological stations: daily observed maximum and minimum air temperature and wind speed. Three kinds of MODIS (Moderate-resolution imaging spectroradiometer) products were used as the inputs for the SEBAL (the Surface Energy Balance Algorithm for Land model) model to estimate the actual evapotranspiration (ET_a) of winter wheat: (i) MOD11A1 (land surface temperature/surface emissivity), (ii) MOD13A2 (NDVI), and (iii) MCD43B3 (surface albedo). For land surface temperature images, cloudy areas were eliminated by replacing the values with the average of two images from the nearest clear dates (Cai and Sharma, 2010; Yang et al., 2015).

8.2.3 CWP estimation

The crop water productivity is defined as the output of crops per unit of water consumption (Perry et al., 2009), as determined in the following equation (Yan and Wu, 2014):

$$CWP = \frac{Y}{ET_a} \qquad \text{Formula 8-1}$$

where, Y is the wheat yield (t·ha^{-1}), ET_a is the corresponding water consumption (mm), and CWP is the winter productivity (kg·m^{-3}).

It differs from water efficiency, which is the ratio of crop yields to the actual amount of water applied to the field (losses included).

The physical location of winter wheat planting has to be known for crop-specific water consumption and water productivity analysis in this study (Cai and Sharma, 2010; Yang et al., 2015). To improve the wheat dominance map, ISODATA class identification and spectral matching proposed by Thenkabail et al. (2007) and widely used in China (Hao et al., 2011) were performed using the ground truth information as the input.

With the development of remote sensing technology, the SEBAL model for ET_a estimation has been applied in the Americas (Paul et al., 2013; Trezza, 2002), Europe (Jacob et al., 2002), Africa (Dzikiti et al., 2016; Kongo et al., 2011), and China (Li et al., 2013; Tang et al., 2013; Yang et al., 2015). These studies demonstrate that the SEBAL model is proved to be with a good efficiency and applicability for water consumption estimations in vegetative area, which is the focus of our study (Cai and Sharma, 2010).

In our study, the SEBAL model based on MODIS data were applied to estimate the daily ET. The SEBAL model is based on the energy balance equation described by the following equation:

$$R_n = G + H + \lambda ET \qquad \text{Formula 8-2}$$

where, R_n is the net radiation(W·m^{-2}); G is the soil heat flux(W·m^{-2}); H is the sensible heat flux(W·m^{-2}); and λET is the latent heat flux associated with evapotranspiration (W·m^{-2}). The main parameters such as the net radiation flux on the land surface (R_n), the soil heat flux (G) and the sensible heat flux (H) can be calculated according to literatures published by Bastiaanssen et al. (1998), Yang et al. (2015) and Ju et al. (2016).

Since the evaporative fraction Λ is constant during a day, the daily ET_{24} (mm) can be estimated using the following equations:

$$\Lambda = \frac{\lambda ET}{R_n - G} \qquad \text{Formula 8-3}$$

$$ET_{24} = \frac{\Lambda(R_{24} - G_{24})}{\lambda} \qquad \text{Formula 8-4}$$

where, ET_{24} is the daily evapotranspiration (mm), G_{24} is the daily soil heat flux (W·m^{-2}), and λ is the latent heat of vaporization (MJ·kg^{-1}).

Hence, the SEBAL was validated to be suitable for estimating evapotranspiration of winter wheat in the Huang-Huai-Hai Plain with correlation coefficient of 0.88 between the estimated and measured values from the field data for Yucheng station in Shandong Province (Yang et al., 2015).

In this study, the county district-level wheat yield map from census data with vectors of administrative county was resolved to the pixels size of the MODIS NDVI data with a stepwise multiple regression equation (Ju et al., 2016). The county-level census data included wheat area and grain yield for 347 counties in the Huang-Huai-Hai Plain and were obtained from the Ministry of Agriculture of China. Areas where wheat was not the dominant crop were masked out using the wheat planting map. The stepwise multiple regression equation was obtained from county district average NDVI values for wheat from the MODIS images and the related district average wheat yield. We obtained the following the linear regression equation was obtained, as shown:

$$\text{Yield}_p = \text{Yield}_{avg} \times \frac{\text{NDVI}_p}{\text{NDVI}_{avg}} \qquad \text{Formula 8-5}$$

where, Yield_p and Yield_{avg} are the average yields of the individual pixel and of the county; NDVI_{avg} is the county-level averaged NDVI; and NDVI_p is the NDVI of individual pixel during the wheat growth stage. The equation was then applied to each pixel on the NDVI wheat subset, leading to a yield map of wheat with 250 m × 250 m resolution. This downscaled method is described in detail in Cai and Sharma (2010).

8.3 Results

8.3.1 ET map

The ET_a of winter wheat was calculated based on wheat-dominant planting map and phenological data for wheat in different agricultural sub-regions. The wheat ET_a map and histogram are shown in Figures 8-1 and 8-2. The seasonal average wheat ET_a from October to the following June are 630 mm (169–729 mm, standard deviation (SD)=49.2 mm) in 2001, 550 mm (203–729 mm, SD=66.2 mm) in 2006, and 538 mm (164–727 mm, SD=84.7 mm) in 2011 (1% of the points were filtered). The average value is detected to be pronouncedly less than the potential wheat ET (680 mm) (Yang et al., 2013a). Figure 8-2 shows the histogram of the average wheat ET_a values, which peak at 641 mm, 574 mm, and 559 mm, respectively, in 2001, 2006, and 2011. The histogram shows that the ET_a range is 320–700 mm, accounting for 99% of the total pixels. The pixels with ET_a values below 320 mm probably represent data collected in the margin of the winter wheat production area.

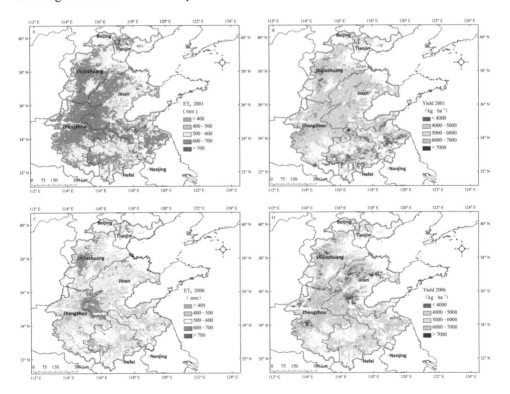

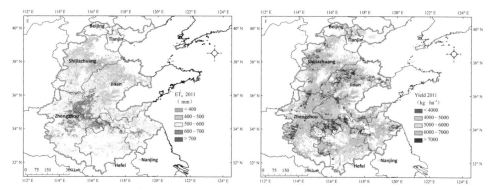

Figure 8-1 Actual evapotranspiration (ET_a, mm) and average yield (Y, ton·ha^{-1}) of winter wheat in the Huang-Huai-Hai Plain (2001, 2006 and 2011). The left maps were obtained from the results based on SEBAL model. The right maps were obtained from grid-level yield with ArcGIS software

In 2001 and 2006, the maximum ET_a for winter wheat in the basin-irrigable land and dry land (Zone 5) and hill-wet hot paddy field (Zone 6) of the Huang-Huai-Hai Plain was 729 mm, while the minimum value were 169 mm and 203 mm in the coastal land-farming-fishing area (Zone 1) and low plain-hydropenia irrigable land and dry land (Zone 2), respectively. A higher ET_a bar between 600 mm and 700 mm

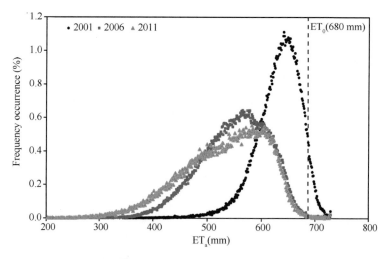

Figure 8-2 The distribution histogram of wheat cropland ET_a in the Huang-Huai-Hai Plain for the period from October to following June in 2001, 2006 and 2011. X-axis represents ET_a (mm) for winter wheat growing season and Y-axis represents frequency occurence (%) for ET_a during winter wheat growing season across the Huang-Huai-Hai Plain. The vertical dashed line represents area-averaged value of ET_0 during winter wheat growing season across the Huang-Huai-Hai Plain according to Yang et al. (2013a)

and then 500 mm and 600 mm was detected in more than 78.0% and 54.8% of the winter wheat area for these two periods, respectively, although significant variations were observed. Furthermore, the maximum ET_a in 2011 was 727 mm in Zone 5, while the minimum value of 164 mm appeared mainly in Zone 1 and Zone 2. We conclude that, compared with 2001, the decreasing areas with an average reduction of 88.4 mm are found in the southern Hebei Province, Henan Province, and northern Jiangsu Province in 2006. Similarly, the decreasing areas with an average reduction of 75.8 mm are found in the Hebei and northern Shandong provinces in 2011 compared with 2006.

8.3.2 Wheat yield map

The wheat yield was in consequence rasterized based on the wheat-dominant planting map and MODIS NDVI products. The wheat yield map is shown in Figure 8-1. The average wheat yield is 5.0 t·ha^{-1}, 5.7 t·ha^{-1}, and 6.3 t·ha^{-1}, ranging from 1.9 t·ha^{-1} to 7.8 t·ha^{-1}, 3.4 t·ha^{-1} to 7.9 t·ha^{-1}, and 3.9 t·ha^{-1} to 8.1 t·ha^{-1} with a standard deviation of 0.08 t·ha^{-1}, 0.07 t·ha^{-1}, and 0.05 t·ha^{-1} respectively (1% of the points were sieved), from October to the following June in 2001, 2006, and 2011.

In 2001 and 2006, the highest wheat yields were more than 7 t·ha^{-1} in Zone 2 and the piedmont plain-irrigable land (Zone 3) of the 3H Plain, while the lowest values were less 1.9 t·ha^{-1} and 3.4 t·ha^{-1} in Zone 6, respectively. A yield belt between 5 t·ha^{-1} and 7 t·ha^{-1} was detected in more than 48.2% and 77.9% of the winter wheat area during these 2001 and 2006, respectively, although significant variations in this value were observed. The highest yield in 2011 was detected in Zones 3 and 5 (more than 8 t·ha^{-1}), while the minimum value of less 3.9 t·ha^{-1} was detected in Zone 1. The pixels with wheat yield values below 1.9 t·ha^{-1}, 3.4 t·ha^{-1}, and 3.9 t·ha^{-1} are probably associated with areas in which the wheat failed due to drought disasters, pest attacks, or unknown causes. Compared with 2001, the 2006 yield increase (average increment of 0.10 t·ha^{-1}), occurred in Shandong, Henan, and northern Jiangsu provinces. Similarly, in 2011 compared with 2006, wheat yield increased, with an average increase of 0.08 t·ha^{-1}, in Henan and northern Anhui provinces.

8.3.3 CWP map

Based on Formula 8-5, the ET and yield data estimated based on remote sensing were used to calculate the annual CWP of winter wheat for 2001, 2006, and 2011 for each pixel in the study area. The wheat WP map and the distribution histogram are shown in Figures 8-3 and 8-4. The results indicate that the average CWP of the plain was 0.79 kg·m^{-3}, 0.99 kg·m^{-3}, and 1.08 kg·m^{-3}, ranging from 0.24 kg·m^{-3} to 1.92 kg·m^{-3}, 0.40 kg·m^{-3} to 1.78 kg·m^{-3}, and 0.50 kg·m^{-3} to 1.99 kg·m^{-3} with a standard deviation of 0.48 kg·m^{-3}, 0.52 kg·m^{-3}, and 0.43 kg·m^{-3} (1% of the points were sieved), respectively, from October to the following June in 2001, 2006, and 2011. The average value in 2001 and 2006 is smaller than the wheat WP (1.05) reported for the Hai Basin (Yan and Wu, 2014). Figure 8-4 shows the histogram of the average wheat WP with a sharp peak of 0.75 kg·m^{-3}, 1.05 kg·m^{-3}, and 1.0 kg·m^{-3}, respectively in 2001, 2006, and 2011.The histogram indicates that the CWP range is 0.4–1.9 kg·m^{-3}, accounting for 99% of the total pixels. The pixels with WP values below 0.4 kg·m^{-3} probably represent data collected in the margin of the winter wheat production area, in misclassified winter wheat areas, or in areas in which the crop failed due to drought disasters, pest attacks, or other causes.

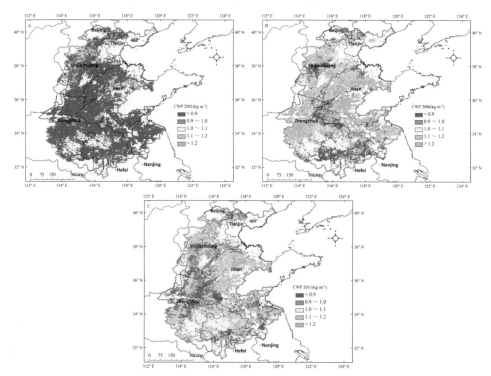

Figure 8-3 Water productivity (WP, kg·m^{-3}) of winter wheat in the Huang-Huai-Hai Plain (2001, 2006 and 2011). These maps were obtained from the results of ET$_a$ and grid-level yield of winter wheat according to the method described as Formula 8-5 with ArcGIS software

The highest WP in 2001 was more than 1.0 kg·m^{-3}, appearing in Zones 2 and 3 of the Huang-Huai-Hai Plain, while the lowest value was 0.5 kg·m^{-3} in Zone 6. The highest value in 2006 and 2011 was more than 1.2 kg·m^{-3} in Zones 4 and 5, while the lowest value was less than 0.9 kg·m^{-3}, mainly in the eastern part of Zones 5, 2, and 6. Furthermore, a higher WP belt between 1.1 kg·m^{-3} and 1.2 kg·m^{-3} was detected in more than 13.8% and 18.7% of the winter wheat area of Zone 4 during these two

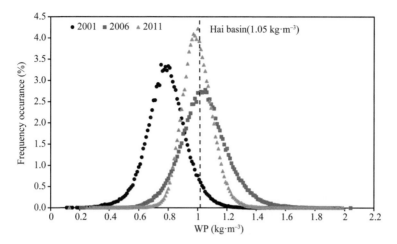

Figure 8-4 The distribution histogram of wheat water productivity in the Huang-Huai-Hai Plain for the period of from October to following June, 2001, 2006 and 2011. X-axis represents crop water productivity (kg·m^{-3}) for winter wheat growing season and Y-axis represents frequency occurrence (%) for crop water productivity during winter wheat growing season across the Huang-Huai-Hai Plain. The vertical dashed line represents the area-averaged value of crop water productivity during winter wheat growing season across the Hai Basin according to Yan and Wu (2014) which is part of Huang-Huai-Hai Plain

periods, respectively, although significant variations of this value were observed. Accordingly, compared with 2001, the increasing area with an average increment of 0.24 kg·m^{-3} was detected mainly in the Shandong, Henan, and northern Jiangsu Provinces in 2006. Similarly, the increasing area with an average increment of 0.11 kg·m^{-3} includes Zones 2 and 5 in 2011 compared with 2006.

8.3.4 Relations among yield, ET$_a$ and CWP

To determine the relations among yield, ET$_a$, and CWP in the Huang-Huai-Hai Plain, correlation analysis was conducted on 0.01°×0.01° pixels in three study periods. Figures 8-5 and 8-6 illustrate the change in the CWP with cumulative ET$_a$ and yield for winter wheat using average values for ET$_a$ and CWP from 2001, 2006, and 2011 across the Huang-Huai-Hai Plain. It is noteworthy that the ET$_a$ data below 200 mm and yield values below 2.0 t·ha^{-1} are not included in the graph.

Figure 8-5 indicates that the CWP decreases with increasing ET$_a$ from 200–750 mm. The relation between the CWP and ET$_a$ has a significance level of 0.05 ($R^2 > 0.12$, n = 4600, 890) for the three studied periods in Zones 5 and 6 of the Huang-Huai-Hai Plain. Furthermore, the correlation coefficient between CWP and ET$_a$ is higher in 2006 than in 2011. However, the relation between CWP and ET$_a$ of the other four zones in 2006 has a significance level of 0.05. The correlation coefficient between CWP and ET$_a$ is higher in 2006 than in 2011. Figure 8-6 indicates that the CWP increases with increasing yield from 2.0–9.0 t·ha^{-1}. The relation between the CWP and yield is pronounced, with a significance level of 0.05 ($R^2 > 0.21$, $n > 4600$, 890) in the three periods in the five agricultural sub-regions, but not in Zone 2. The correlation coefficient between the CWP and yield is higher in 2011 than in 2006. Hence, the wheat yield has a closer linear relationship with CWP than with ET_a in the Huang-Huai-Hai Plain.

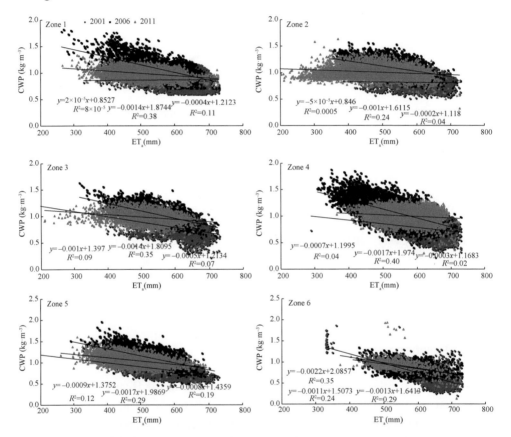

Figure 8-5 Relationships between actual evapotranspiration (ET_a) and crop water productivity for winter wheat in the Huang-Huai-Hai Plain, using average values of ET_a and crop water productivity for 2001, 2006 and 2011 across the plain. X-axis is area-averaged ET_a value with the unit of mm, and the left Y-axis is area-averaged crop water productivity of winter wheat with the unit of kg·m^{-3}

Based on the combined correlation analysis described in Figures 8-5 and 8-6, we conclude that the Yield increase principally controls the increase of the water productivity in Zones 1 and 4; the increase of the water productivity is governed more by the increment of the yield than by the reduction of ET_a in the other four agricultural sub-regions.

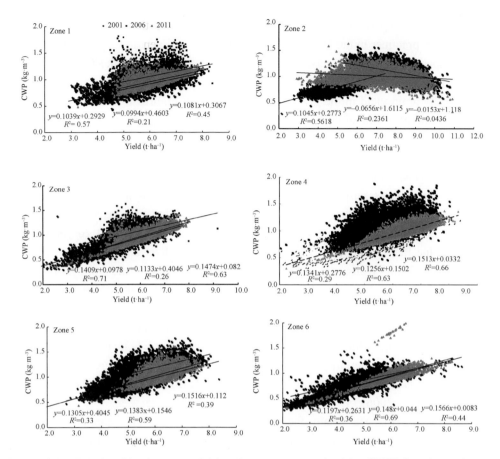

Figure 8-6 Relationships between yield and crop water productivity (CWP) for winter wheat in the Huang-Huai-Hai Plain, using average values for yield and crop water productivity for 2001, 2006 and 2011 across the plain. X-axis is area-averaged yield of winter wheat with the unit of t·ha^{-1} and the left Y-axis is area-averaged crop water productivity of winter wheat with the unit of kg·m^{-3}

8.4 Discussion

The range of the CWP of winter wheat in the Huang-Huai-Hai Plain (0.24 kg·m^{-3} to 1.99 kg·m^{-3}) for three periods, 2001, 2006, and 2011, is in accordance with results simulated for the Northern Plain, China (Chao et al., 2009), and results estimated from remote sensing data (Liu et al., 2010a) but lower compared with experimental data from the Luancheng Station (Zhang et al., 2011). It is not unexpected that the CWP is higher at the research stations because researchers perform studies in a given condition to achieve higher potential yield per hectare and have more resources than regular farmers (Yan and Wu, 2014; Zhang et al., 2011). The relatively low values for the

CWP in the Huang-Huai-Hai Plain suggest that there is room for improvement for the agricultural practices in the Huang-Huai-Hai Plain.

The spatial analysis demonstrates the linear relation between the CWP, ET_a, and grain yield across the Huang-Huai-Hai Plain. As a whole, the CWP values are more affected by grain yield in the given range described in Figures 8-5 and 8-6. The CWP increase are observed along with the increment of the grain yield rather than the reduction of ET in accordance with reported results in the Indo-Gangetic River Basin (Cai and Sharma, 2010) and Hai Basin (Yan and Wu, 2014). Dong et al. (2007) reported a closer linear relationship between the CWP and grain yield under different irrigation treatments, despite much differences of the CWP among 19 wheat species in the North China Plain. Zhang et al. (2005) also observed that an increase of wheat yield by 50% and consequently significant CWP improvements in the past 20 years in the North China Plain. The yield increase is due to the fertilizer and pesticide use in the North China Plain (Wang et al., 2010). The relationship between the grain yield and ET in the six agricultural sub-regions of the Huang-Huai-Hai Plain in 2011 is described by a shape of typical parabola. The wheat yield increases before reaching the parabola peak (the yield is 5.8 t·ha^{-1} and ET_a is 522 mm in the northern zones; the yield is 5.9 t·ha^{-1} and ET_a is 460 mm in the southern zones) and then drops as the ET_a increases. This can be concluded that a higher irrigation amount does not inevitably achieve higher grain yield, which is similar to results of field experiments from other researchers (Li et al., 2005; Sun et al., 2006).

The Huang-Huai-Hai Plain faces a serious water shortage threat due to the excessive exploitation of groundwater and precipitation reductions at an average rate of 2.92 mm·y^{-1} (Liu et al., 2010b). In recent years, many studies have focused on the crop water use in the Huang-Huai-Hai Plain (Li et al., 2008; Yang et al., 2013b). However, irrigation was widely accepted to be the dominant water resource in the Hebei and Henan Provinces and 60% of total water consumption depends on it. The winter wheat production in the Huang-Huai-Hai Plain strongly depends on supplementary irrigation (Li et al., 2008; Sun et al., 2006). According to Yang et al. (2013a), field management is a more important factor with respect to ET_a of winter wheat compared with climate transitions and geography. Although the water saving method is promoted in the Huang-Huai-Hai Plain, traditional surface-flooding irrigation is still common in the irrigation district. The irrigation efficiency in Northern China was 0.42 in 2010, and much lower than the world average (Bai et al., 2011). Liu et al. (2011) hold the view that the relatively higher grain yield and CWP of winter wheat could be gained through a sprinkler irrigation in amount of 63% ET_a in winter wheat production area.

It is widely accepted that the improvement of the CWP is more complicated on a plain scale than in field experiments due to factors such as crop cultivar, soil type, drought disaster, and pest attacks. China has a good extension system to promote suitable technology for farmers. Deng et al. (2006) found that mulching with crop residues could increase the CWP by 10%–20% by reducing soil evaporation together with increasing plant transpiration. Straw mulching can easily be implemented and extended in the Huang-Huai-Hai Plain. In addition to droughts, the soil infertility is a factor constraining the productivity. A 15-year field experiment indicated that nitrogen fertilizer is important for the improvement of the CWP, and phosphate fertilizer is highly effective in increasing the total soil water use at the Changwu Ecological Station (Dang, 1999). Furthermore, the winter wheat cultivars have been changed five times in the past 20 years, with the average yield increasing due to the increase of the average kernel numbers per spike (Zhou et al., 2007). Other land and crop interventions, such as drought adaption, insect control, are also important for yield improvement in addition to water management.

8.5 Conclusions

The average CWP of winter wheat in the Huang-Huai-Hai Plain is in the range of 0.24 kg·m^{-3} to 1.99 kg·m^{-3} in 2001, 2006 and 2011. The spatial analysis demonstrates that the CWP values are more affected by the grain yield based on the linear relation between CWP and ET$_a$ as well as the grain yield across the Huang-Huai-Hai Plain. Furthermore, the grain yield increases to a critical value of 522 mm and 460 mm in the northern and southern zones, respectively, and then drops as the ET$_a$ increases. The temporal analysis shows an average increase of 0.25 kg·m^{-3} and 0.09 t·ha^{-1} of CWP and yield, respectively, and an average reduction of 82.1 mm for ET$_a$ during the studied periods.

The relatively low CWP across the plain suggests that it is possible to achieve a higher yield for the same amount of water consumption in the plain, which should be a goal for agricultural investments in water-scarce regions. In general, the findings presented in this paper suggest that the basic equation describing the relationship between CWP, ET$_a$, and grain yield imply that improvement of CWP depends on the ET$_a$ reduction and, more specifically, evaporation. However, the analysis results show that this measure did not play a substantial role in improving the CWP so far. Consequently, adaptive measures, such as cultivar alternatives, disaster and pest control, supplementary irrigation scheduling, and residue mulching, ought to be adopted by farmers to achieve a CWP improvement in the six agricultural sub-regions of the Huang-Huai-Hai Plain.

References

Ali M H, Talukder M S U. 2008. Increasing water productivity in crop production: A synthesis.

Agricultural Water Management, 95(11): 1201-1213.

Bai G, Zhang R, Geng G, Ren Z, Zhang P, Shi J. 2011. Integrating agricultural water-saving technologies in Hetao Irrigation District. Bulletin of Soil and Water Conservation, 31(1): 149-154.

Bastiaanssen W G M, Menenti M, Feddes R A, Holtslag A A M. 1998. The surface energy balance algorithm for land (SEBAL): Part 1 formulation. Journal of Hydrology, 212(98): 801-811.

Boutraa T. 2010. Improvement of water use efficiency in irrigated agriculture: A review. Journal of Agronomy, 9(1): 1-8.

Cai X L, Sharma B R. 2010. Integrating remote sensing, census and weather data for an assessment of rice yield, water consumption and water productivity in the Indo-Gangetic river basin. Agricultural Water Management, 97(2): 309-316.

Cai X, Molden D, Mainuddin M, Sharma B, Ahmad M U D, Karimi P. 2011. Producing more food with less water in a changing world: Assessment of water productivity in 10 major river basins. Water International, 36(1): 42-62.

Cao X, Wang Y, Wu P, Zhao X. 2015. Water productivity evaluation for grain crops in irrigated regions of China. Ecological Indicators, 55: 107-117.

Carruthers I, Rosegrant M W, Seckler D. 1997. Irrigation and food security in the 21st century. Irrigation & Drainage Systems, 11(2): 83-101.

Chao C, Qiang Y, En-li W, Jun X. 2009. Modeling the spatial distribution of crop water productivity in the North China Plain. Resources Science, 31(9): 1477-1485.

Dang T. 1999. Effects of fertilization on water use efficiency of winter wheat in arid highland. Eco-Agricultural Research, 7(2): 28-31.

Deng X, Shan L, Zhang H, Turner N C. 2006. Improving agricultural water use efficiency in arid and semiarid areas of China. Agricultural Water Management, 80(1): 23-40.

Dong B D, Zhang Z B, Liu M Y, Zhang Y Z, Li Q Q, Shi L, Zhou Y T. 2007. Water use characteristics of different wheat varieties and their responses to different irrigation schedulings. Transactions of the Chinese Society of Agricultural Engineering, 23(9): 27-33.

Dzikiti S, Gush M B, Le Maitre D C, Maherry A, Jovanovic N Z, Ramoelo A, Cho M A. 2016. Quantifying potential water savings from clearing invasive alien Eucalyptus camaldulensis using in situ and high resolution remote sensing data in the Berg River Catchment, Western Cape, South Africa. Forest Ecology & Management, 361: 69-80.

Hao W, Mei X, Cai X, Du J, Liu Q. 2011. Crop planting extraction based on multi-temporal remote sensing data in Northeast China. Nongye Gongcheng Xuebao/transactions of the Chinese Society of Agricultural Engineering, 27(1): 201-207.

Huang F, Li B. 2010. Assessing grain crop water productivity of China using a hydro-model-coupled-statistics approach. Part II: Application in breadbasket basins of China. Agricultural Water Management, 97(9): 1259-1268.

Igbadun H E, Mahoo H F, Tarimo A K, Salim B A. 2006. Crop water productivity of an irrigated maize crop in Mkoji sub-catchment of the Great Ruaha River Basin, Tanzania. Agricultural Water Management, 85(1): 141-150.

Immerzeel W W, van Beek L P, Bierkens M F. 2010. Climate change will affect the Asian water towers. Science, 328(5984): 1382-1385.

Jacob F, Olioso A, Gu X F, Su Z, Seguin B. 2002. Mapping surface fluxes using airborne visible, near infrared, thermal infrared remote sensing data and a spatialized surface energy balance model. Agronomie, 22(6): 669-680.

Jia J, Liu C. 2002. Groundwater dynamic drift and response to different exploitation in the North China Plain: A case study of Luancheng County, Hebei Province. Acta Geographica Sinica-Chinese Edition, 57(2): 201-209.

Ju H, Liu Q, Yang J Y, Yan C. 2016. Potential Effect of Climate Drought on the Yield and Water Productivity of Winter Wheat over the Huang-Huai-Hai Plain. Beijing: Science Press.

Kongo M V, Jewitt G W P, Lorentz S A. 2011. Evaporative water use of different land uses in the upper-Thukela river basin assessed from satellite imagery. Agricultural Water Management, 98(11): 1727-1739.

Li F, Li Q, Wang Z. 2009. Analysis of Agricultural Water Saving in Hai Basin Based on Remote Sensing Water Productivity, Proceedings of International symposium of Hai Basin Integrated Water and environment management. Beijing: Orient Academic Forum: 413-418.

Li H, Zheng L, Lei Y, Li C, Liu Z, Zhang S. 2008. Estimation of water consumption and crop water productivity of winter wheat in North China Plain using remote sensing technology. Agricultural Water Management, 95(11): 1271-1278.

Li J, Inanaga S, Li Z, Eneji A E. 2005. Optimizing irrigation scheduling for winter wheat in the North China Plain. Agricultural Water Management, 76(1): 8-23.

Li Q Q, Zhou X B, Chen Y H, Yu S L. 2012. Water consumption characteristics of winter wheat grown using different planting patterns and deficit irrigation regime. Agricultural Water Management, 105: 8-12.

Li Z, Liu X, Ma T, Kejia D, Zhou Q, Yao B, Niu T. 2013. Retrieval of the surface evapotranspiration patterns in the alpine grassland-wetland ecosystem applying SEBAL model in the source region of the Yellow River, China. Ecological Modelling, 270: 64-75.

Liang W L, Carberry P, Wang G Y, Lü R H, Lü H Z, Xia A P. 2011. Quantifying the yield gap in wheat-maize cropping systems of the Hebei Plain, China. Field Crops Research, 124(2): 180-185.

Liu C, Shi R, Gao W, Gao Z. 2010a. Analyze the regional water budget in Shandong province by applying the evapotranspiration remote sensing method. Journal of Natural Resources, 25(11): 1938-1948.

Liu H, Yu L, Luo Y, Wang X, Huang G. 2011. Responses of winter wheat (*Triticum aestivum* L.) evapotranspiration and yield to sprinkler irrigation regimes. Agricultural Water Management, 98(4): 483-492.

Liu J, Wiberg D, Zehnder A J, Yang H. 2007. Modeling the role of irrigation in winter wheat yield, crop water productivity, and production in China. Irrigation Science, 26(1): 21-33.

Liu S, Mo X, Lin Z, Xu Y, Ji J, Wen G, Richey J. 2010b. Crop yield responses to climate change in the Huang-Huai-Hai Plain of China. Agricultural Water Management, 97(8): 1195-1209.

Molden D, Oweis T Y, Pasquale S, Kijne J W, Hanjra M A, Bindraban P S, Bouman B A, Cook S, Erenstein O, Farahani H, Hachum A. 2007. Pathways for increasing agricultural water productivity. *In*: Molden D. Water for Food, Water for Life. A Comprehensive Assessment of Water Management in Agriculture. London: Springer, 21: 278-310.

Molden D, Oweis T, Steduto P, Bindraban P, Hanjra M A, Kijne J. 2010. Improving agricultural water productivity: Between optimism and caution. Agricultural Water Management, 97(4): 528-535.

Murray S, Foster P, Prentice I. 2012. Future global water resources with respect to climate change and water withdrawals as estimated by a dynamic global vegetation model. Journal of Hydrology, 448: 14-29.

Paul G, Gowda P H, Prasad P V, Howell T A, Aiken R M, Neale C M. 2013. Investigating the influence of roughness length for heat transport (zoh) on the performance of SEBAL in semi-arid irrigated and dryland agricultural systems. Journal of Hydrology, 509(4): 231-244.

Perry C, Steduto P, Allen R G, Burt C M. 2009. Increasing productivity in irrigated agriculture: Agronomic constraints and hydrological realities. Agricultural Water Management, 96(11): 1517-1524.

Perry C. 2011. Accounting for water use: Terminology and implications for saving water and increasing production. Agricultural Water Management, 98(12): 1840-1846.

Playán E, Mateos L. 2006. Modernization and optimization of irrigation systems to increase water productivity. Agricultural Water Management, 80(1-3): 100-116.

Ren J, Chen Z, Zhou Q, Tang H. 2008. Regional yield estimation for winter wheat with MODIS-NDVI data in Shandong, China. International Journal of Applied Earth Observation and Geoinformation, 10(4): 403-413.

Rosegrant M, Cai X, Cline S A. 2002. World water and food to 2025. International Food Policy Research Institute, Washington, DC. doi:0-89629-646-6.

Sun H, Liu C, Zhang X, Shen Y, Zhang Y. 2006. Effects of irrigation on water balance, yield and WUE of winter wheat in the North China Plain. Agricultural Water Management, 85(1): 211-218.

Sun Q, Kröbel R, Müller T, Römheld V, Cui Z, Zhang F, Chen X. 2011.Optimization of yield and water-use of different cropping systems for sustainable groundwater use in North China Plain. Agricultural Water Management, 98(5): 808-814.

Tang R, Li Z L, Chen K S, Jia Y, Li C, Sun X. 2013. Spatial-scale effect on the SEBAL model for evapotranspiration estimation using remote sensing data. Agricultural & Forest Meteorology, 174-175: 28-42.

Thenkabail P S, Gangadhararao P, Biggs T W, Krishna M, Turral H. 2007. Spectral matching techniques to determine historical land-use/land-cover (LULC) and irrigated areas using Time-series 0.1-degree AVHRR pathfinder dataset. Photogrammetric Engineering & Remote Sensing, 73(9): 1029-1040.

Trezza R. 2002. Evapotranspiration using a satellite-based surface energy balance with standardized ground control. Ph. D. dissertation, USU, Logan, UT, 339.

Vrsmarty C J, Green P, Salisbury J, Lammers R B. 2000. Global water resources: Vulnerability from climate change and population growth. Science, 289(5477): 284-288.

Wallensten A, Munster V J, Latorre-Margalef N, Brytting M, Elmberg J, Fouchier R A, Fransson T, Haemig P D, Karlsson M, Lundkvist A, Osterhaus A D. 2007. Surveillance of influenza A virus in migratory waterfowl in northern Europe. Emerging Infectious Diseases, 13(3): 404-411.

Wang S, Song X, Wang Q, Xiao G, Liu C, Liu J. 2009. Shallow groundwater dynamics in North China Plain. Journal of Geographical Sciences, 19(2): 175-188.

Wang X, Li Z, Ma W, Zhang F S. 2010. Effects of fertilization on yield increase of wheat in different agro-ecological regions of China. Scientia Agricultura Sinica, 43(12): 2469-2476.

Wolf A T, Yoffe S B, Giordano M. 2003. International waters: Identifying basins at risk. Water Policy, 5(1): 29-60.

Yan N, Wu B. 2014. Integrated spatial-temporal analysis of crop water productivity of winter wheat in Hai Basin. Agricultural Water Management, 133: 24-33.

Yang J Y, Mei X R, Huo Z G, Yan C R, Ju H, Zhao F H, Qin L. 2015. Water consumption in summer maize and winter wheat cropping system based on SEBAL model in Huang-Huai-Hai Plain, China. Journal of Integrative Agriculture, 14(10): 2065-2076.

Yang J Y, Qin L, Mei X R, Yan C R, Ju H, Xu J W. 2013a. Spatiotemporal characteristics of reference evapotranspiration and its sensitivity coefficients to climate factors in Huang-Huai-Hai Plain, China. Journal of Integrative Agriculture, 12(12): 2280-2291.

Yang X, Gao W, Shi Q, Chen F, Chu Q. 2013b. Impact of climate change on the water requirement of summer maize in the Huang-Huai-Hai farming region. Agricultural Water Management, 124: 20-27.

Yoffe S, Wolf A T, Giordano M. 2003. Conflict and cooperation over international freshwater resources: Indicators of basins at risk. Journal of the American Water Resources Association, 39(5): 1109-1126.

Zhang X, Chen S, Liu M, Pei D, Sun H. 2005. Improved water use efficiency associated with cultivars and agronomic management in the North China Plain. Agronomy Journal, 97(3): 783-790.

Zhang X, Chen S, Sun H, Shao L, Wang Y. 2011. Changes in evapotranspiration over irrigated winter wheat and maize in North China Plain over three decades. Agricultural Water Management, 98(6): 1097-1104.

Zhao R F, Chen X P, Zhang F S, Zhang H, Schroder J, Römheld V. 2006. Fertilization and nitrogen balance in a wheat-maize rotation system in North China. Agronomy Journal, 98(4): 938-945.

Zhou Y, He Z H, Sui X X, Xia X C, Zhang X K, Zhang G S. 2007. Genetic improvement of grain yield and associated traits in the Northern China winter wheat region from 1960 to 2000. Crop Science, 47(1): 245-253.

Zwart S J, Bastiaanssen W G. 2004. Review of measured crop water productivity values for irrigated wheat, rice, cotton and maize. Agricultural Water Management, 69(2): 115-133.

Chapter 9　General discussion, conclusions, and prospects

Abstract

Drought and water shortage are generally acknowledged to be one of the most critical problems faced by the agricultural production in the Huang-Huai-Hai Plain. The overall ambition of this book was to investigate the extent to which the yield and crop water productivity for winter wheat could respond to climate change and drought associated with their improvements using SPEI-PM method, DSSAT-CERES-Wheat model, SEBAL model and remote sensing data across the Huang-Huai-Hai Plain, which is the wheat production base of China. This chapter first summarizes the main results and how they support the hypotheses raised throughout this book, and then provides a discussion of the main findings. Finally, we conclude with implications of this research followed by an outlook for future research and a broader perspective about the role of scientific knowledge in informing policy and decision making. Our results can have major agronomic consequences regarding the reform of the common agricultural policy in the Huang-Huai-Hai Plain, China. It will be adopted to develop feasible straw (film) mulching, regulated deficit irrigation, and soil water storage and preservation to reduce pressure on groundwater over-exploitation, especially for winter wheat in the Huang-Huai-Hai Plain. The results are expected to provide basis information for agricultural water management, improvement of crop water productivity and choice of adaptive mechanism under climate change in the Huang-Huai-Hai Plain. Finally, it is worth mentioning that our results have some uncertainties due to lack of data of deficit irrigation and fertilizer practices for CERES-Wheat calibration, and observed crop water productivity in agro-meteorological stations for temporal analysis, and the uncertainties arising from the input data of CERES-Wheat modeling.

9.1 Overview of results and hypotheses

Drought and water shortage are generally accepted to be one of the most critical problems faced by worldwide agriculture, and it is especially so in China where agricultural production is largely dependent on the timely, adequate, and proper distribution of rainfall. The overall ambition of this book was to investigate the extent to which the grain yield and crop water productivity for winter wheat can respond to climate change and drought across the Huang-Huai-Hai Plain, which is the wheat production base of China. This chapter first summarizes the main results and how they support the hypotheses raised throughout this book, and then provides a discussion of the main findings. Finally, we conclude with implications of this research followed by an outlook for future research and a broader perspective about the role of scientific knowledge in informing policy and decision making.

9.1.1 ET_0 and drought characteristics

In our study, the temporal characteristics in ET_0 and its response to climatic variables were witnessed using Mann-Kendall test and a partial derivative method based on a combined dataset composed of a historical 54-year time series and the RCP8.5 scenario. Subsequently, the drought characteristics were investigated for 1961–2010 and future 2010–2099 under RCP8.5 climate scenario based on the evaluation of drought indices over the Huang-Huai-Hai Plain, China.

SPEI based on FAO-56 Penman-Monteith formula is recognized to be more suitable for agricultural drought impact analysis over the Huang-Huai-Hai Plain due to its higher correlation coefficients with historical drought data. However, it does not mean that SPEI-PM would have the same applicability in other locations or systems. Remarkably, the best drought index for detecting impacts changes as a function of the analyzed system and the performance of the drought indices varied spatially (Vicente-Serrano et al., 2012). For example, the SPI is found to be well correlated with runoff anomaly in 10 large regions of China (Zhai et al., 2010), while SPEI is better for hydrological application in western Canada. Additionally, the comparison between SPEI-PM and SPEI-TH indicated that the way which ET_0 is estimated would make differences in drought applicability and long-term drought trend. This difference has also been found in other places of China (Wang et al., 2015b; Xu et al., 2015; Zhang et al., 2015) and around the world (Begueria et al., 2014; Sheffield et al., 2012). Both Thornthwaite (TH) and Penman-Monteith (PM) are widely used in drought index estimation, while the TH model used for computing ET_0 in drought assessment is popularly used due to its simplicity and less data requirements (only temperature), such as in original SPEI and PDSI indices. Chen et al. (2005) concluded that the TH method overestimates ET_0 in southeast China where ET_0 is low, and underestimates in the northern and northwest parts where ET_0 is high when compared to pan data, and it does not follow the temporal variation well. Instead, PM equation is the most reliable estimation and recommended by the FAO to calculate crop water requirements (Allen et al., 2005).

Total drought events in the Huang-Huai-Hai Plain were described as no significant tendency over the historical period due to the reduction of drought duration, severity, and intensity by the decreasing ET_0. In contrast with the historical period, drought characteristics trends to be lower in the first thirty years in the future RCP8.5 scenario; Nevertheless it is predicted to be intensified in the 2055 and the 2085. The precipitation is expected to increase by 1.88 mm·y^{-1} in the future period, but the amplification of ET_0 by PM equations counteracted this increase. Thus, SPEI-PM predicted that almost all meteorological stations would experience significant drying trends, except those in the southwest regions where SPEI-PM described an insignificant trend. Whereas ET_0 was detected to slightly decrease in the past 54 years, the projected scenario RCP8.5 describes a bigger increment of 1.36 mm·y^{-1} in summer and 3.37 mm·y^{-1} in the entire year. Results from Chapter 2 and 3 confirm the hypothesis that drought conditions will aggravate due to increasing ET_0 induced by climate change and augment crop water consumption in studied region.

9.1.2 Effects of climate change and drought on wheat yield

The issue of climate change is one of the most discussed topics among individuals and organizations, especially within the scientific community. The effect of future scenario on crop production has been widely investigated using DSSAT-CERES model (Challinor and Wheeler, 2008; Guo et al., 2010; Tao et al., 2008a). The relative impact of the change in solar radiation, maximum temperature, minimum temperature and precipitation on wheat yield was investigated using CERES-Wheat model for the timespan of the last 30 years (1985–2014) and the next 30 years (2021–2050) under the RCP4.5 and RCP8.5 pathways. As a matter of prime importance, the potential impact of drought related to climate change on wheat yield and the probability distribution of yield reduction were estimated over 12 selected locations in the Huang-Huai-Hai Plain during the last 1981–2015 years.

These chapters aimed to resolve the hypothesis that the response of winter wheat yield to climate change and drought is available throughout the Huang-Huai-Hai Plain. Increases in temperature and precipitation result in positive evolutions: the changes in these variables raise simulated yield by 6.2% and 14.2%, respectively, under RCP8.5 in the Huang-Huai-Hai Plain. The results are supported by the results of Tao et al. (2008b; 2014) and Xiao et al. (2008), who concluded that the wheat yield has benefited from climate warming in northern and northwest China.

Rainfall has always been a limiting factor for the agricultural production in China (Piao et al., 2010), especially in the Huang-Huai-Hai Plain, where only 25%–40% of water demand is satisfied by rainfall during the wheat growing seasons (Mei et al., 2013). Despite the uncertainty in precipitation amounts and the spatial patterns simulated by climate models (Xin et al., 2013), our study showed that the increasing rainfall in the next 30 years could provide important benefits in terms of the wheat yield production to reduce the pressure of groundwater resource. Nevertheless, many other studies have found that there was no significant influence of rainfall on yield based on long-term field observations or statistical datasets (Xiao and Tao, 2014; Xiong et al., 2012; Tao et al., 2014), since the water deficit is compensated by pumped groundwater, and the impact of rainfall changes was smoothed over. To fully reenact the rainfall changes to yield, the wheat growth and development was therefore simulated in our study under rainfall conditions (no irrigation). Thus, the contributions of increasing rainfall to wheat yield were higher than those in other studies. Nevertheless, the supposed positive impact of increasing solar radiation is visible in our simulations. The result is inconsistent with previous studies, which suggest that crop yields are positively correlated with solar radiation, using both processed- crop models (Xiong et al., 2012) and statistical approaches (Tao et al., 2014). Higher solar radiation also increases the evaporation amount and water stress and consequently causes the photosynthetically active radiation conversion to dry matter ratio (PARUE) to decrease. Logically, these negative impacts are lower at southern stations of the Huang-Huai-Hai Plain with less water stress.

The results from the analysis of cumulative probability for grain yield reduction described a significant gap between the observed and potential yields, which can be resulted from water stress and management inputs variations. Furthermore, the grain yield reduction was found to be larger during jointing to heading stage compared to the filling stage, which is in agreement with experimental results in Anhui province (Wang et al., 2001) and simulated findings by WOFOST model in Zhengzhou agro-meteorological station (Zhang et al., 2012). The drought reduced wheat yield by affecting grain-filling intensity due to soil water deficit. The reduction in yield was significantly different when drought occurred at different developmental stages. There was a higher impact on winter wheat when drought occurred at several developmental stages than at a single developmental stage.

9.1.3 Spatial variability in crop water productivity

In our study, an attempt has been made to partition the actual regional evapotranspiration for wheat from the double crop rotation system using SEBAL and crop dominance information. On this basis, We investigated the relations among yield, CWP, and ET to grasp the characteristics of water productivity for winter wheat in the Huang-Huai-Hai Plain.

The grid-level wheat yield was generated from the county-level yield using estimated multiple regression equation based on MODIS NDVI and crop dominance map. In consequence, the averaged water productivity was estimated as 0.95 kg·m^{-3} with CWP values across the plain ranging from 0.24 kg·m^{-3} to 1.99 kg·m^{-3} for three periods of 2001, 2006 and 2011 for winter wheat across the Plain. From the correlation analysis, we can conclude that the yield increase principally controlled increase of water productivity in north agricultural sub-regions and the increase of water productivity was more governed by the increment of yield than the reduction of ET_a in other agricultural sub-regions. Results from this chapters confirm the hypothesis that the relationship between water productivity & ET and yield of winter wheat should be defined at the level of the sub-agricultural regions in the Huang-Huai-Hai Plain.

The averaged seasonal ET_a of winter wheat was 572.7 mm and detected to have a significant relationship with latitude from October of 2001, 2006 and 2011 to following June over the Huang-Huai-Hai Plain. Our study is the first attempt to apply SEBAL model and crop information to estimation of wheat ET_a from double crop rotation at a large regional scale, such as the Huang-Huai-Hai Plain, which manifests performance in our research over diverse previous studies (Cai and Sharma, 2010; Immerzeel et al., 2008; Zwart and Bastiaanssen, 2007; Zwart et al., 2010).

Sustainable irrigation practices and adequate water allocation strategies at the right spatial scale are crucial to avoiding over-exploitation of various resources (Condon and Maxwell, 2014; Esnault et al., 2014). Details of crop water use patterns are needed for water management including irrigation practices and agriculture water use strategies in water-short areas at the spatial level of a particular irrigation region because it is at this level that sustainable water supply for agriculture can meaningfully be improved by active management. Estimating the water balance and especially the total amount of irrigation water applied is a complex task in researches of crop water consumption structure. Current estimates of irrigation water applied in regional scale are mainly based on plot-scale experiments (Chen et al., 2002), water balance methods (Cheema et al., 2014; Ruud et al., 2004), or surveys, which are costly and often associated with practical and legal difficulties. Additionally, such estimates and surveys are only useful for a specific area, and cannot be expanded to large-scale areas, because of difference in farming systems, canal command areas, and especially irrigation schemes (Molden et al., 2003).

9.2 General discussion

9.2.1 Agricultural adaptations for CWP improvements

Over-irrigation analysis on water consumption of winter wheat

In this section, we estimated the effective precipitation deficit (PD_ET$_a$) by subtracting the cumulative precipitation (P) from the cumulative ET$_a$. This approach has been used previously by Pauw and Pauw (2002) in an agro-ecological study of the Arabian Peninsula and in Northern China (Liu et al., 2013). A positive value indicates water is in excess of crop water requirements (ET$_c$) and a negative value indicates a deficit in terms of ET$_c$. Crop water requirement (ET$_c$) is calculated from ET$_0$ as suggested, usually, by Food and Agriculture Organization of the United Nations (FAO). The guidelines suggested methods of ET$_c$ derivation and discussed the application of data on crop water requirements in irrigation project planning, design and operation (Liu et al., 2015). In consequence, the water requirement deficit (ED) was estimated by subtracting the cumulative ET$_c$ from ET$_a$. A positive value indicates water consumption in excess of crop water requirements and a negative value indicates a deficit in terms of crop water requirements. In our study, a correlation analysis of crop water productivity with the precipitation deficit (PD_ET$_a$) and water requirement deficit (ED) was conducted on winter wheat (as an instance) in 2011–2012 over Huang-Huai-Hai Plain. As described in Figure 9-1, supplementary irrigation was needed to get higher yield for winter wheat due to irregular and concentrated precipitation. The values of correlation coefficient demonstrated higher potential for crop water productivity of winter wheat under alleviated water stress. In the same way, the negative correlation of water productivity and ED indicated that crop water productivity of winter wheat deceased in light of increasing ED. The positive value from water consumption in excess of crop water requirements was resulted from unreasonable supplementary irrigation mode. Consequently, adapted irrigation schedules could be defined and strategies could be developed to optimize water use as well as yield. The present study gives a first hint of expected trends and regions to prioritize. Furthermore, results indicated that the grain yield increased before critical value of 522 mm in north zones and 460 mm in south zones and then dropped as the ET$_a$ increased. A study in North China revealed that under wheat production, additional mulching reduced mean daily soil evaporation by 16% and 37%, respectively (Chen et al., 2007). In Spain, Döll (2002) has predicted a declining in irrigation requirements by 2020 on account of the possible early sowing under more favorable higher temperatures.

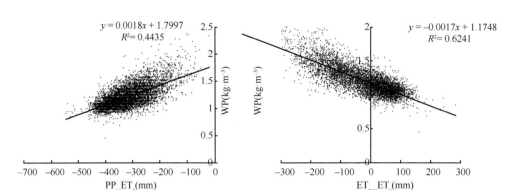

Figure 9-1 The correlation of crop water productivity with the precipitation deficit (PP_ET$_a$), and the water requirement deficit (ET$_a$_ET$_c$) for winter wheat in 2011–2012 over the Huang-Huai-Hai Plain

Adaptive capacity to potential drought

The grain yield damage of winter wheat caused by drought is inclined to accelerate in the future, described by a significantly increasing of drought events under RCP8.5. Li et al. (2015) documented a modeling approach by using crop model DSSAT and hydrological indices to assess the vulnerability of winter wheat to future potential drought, based on an integrated assessment of exposure, sensitivity and adaptive capacity in the Huang-Huai-Hai Plain.

As presented in Figure reported by Li et al. (2015), areas with very high and high grades to adaptive capacity to irrigation are mainly located in Beijing, Tianjin and Hebei, which indicates that the irrigation is more advanced in these regions than in other regions. The beneficial effect of irrigation is more evident in the north Huang-Huai-Hai Plain than in the south. Crop yields under no irrigation condition are relatively high in southern Huang-Huai-Hai region since the southern part generally receives more rainfall than in the north and therefore is less exposed to potential drought. Consequently, under RCP8.5 emission scenario, the worst drought effect is projected to occur around 2030. With increasing drought risks, we suggest immediate and appropriate adaptation actions to be taken before the 2030s, especially in Shandong and Hebei provinces, the most vulnerable provinces of Huang-Huai-Hai Plain. Consequently, adaptive measures such as straw mulching, soil water preservation, irrigation scheduling are encouraged to in agriculture practice to cope with ground water decline especially in winter wheat growing stage over the Huang-Huai-Hai Plain.

Alternative tillage to increase water storage in the soil profile

Increasing CWP and drought tolerance for winter wheat in virtue of genetic improvements and physiological regulation is supposed to be alternatives to achieve efficient water application. Ali and Talukder (2008) synthesized of the factors affecting crop yields and water productivity, and the possible techniques for improving water productivity. Increasing water storage within the soil profile is necessary to increase plant available soil water. Conventionally, tillage can roughen the soil surface and break any soil crust, which in subsequence bring about water storage increased. Li et al. (2007) has already reported that change in bulk density depends on the intensity of tillage systems. Different tillage systems produce different results like No tillage (NT) promotes SOC sequestration (Dick et al., 1991; Liu et al., 2014a) and improved soil aggregates (Lal et al., 1994), while conventional tillage (CT) usually increases available water capacity and infiltration rate and decreases runoff (Wright et al., 1999). A meta-analysis of the yield based on 807 experiments and water productivity based on 501 experiments was conducted for winter wheat under different tillage in China. From the Figure 9-2, it can be seen that NT and deep tillage increased yield of winter wheat by 6% and 9.5% respectively and water productivity by 4.5% and 9.0% respectively compared with CT in China. Adopting a better tillage system improves not only the soil properties and crop productivity but also water productivity. In the Huang-Huai-Hai Plain, deep tillage should be chosen to improve the soil water storage in south part due to precipitation shortage, while NT turns to be a good alternative to promotes SOC sequestration and reduce soil evaporation.

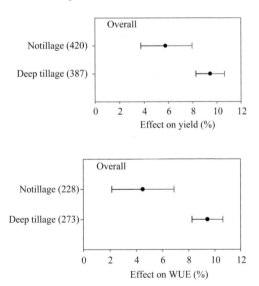

Figure 9-2 A meta-analysis of the yield and crop water productivity for winter wheat under different tillage in China. The top map was obtained from 807 papers 420 of which are about no-tillage and 387 of which are about deep tillage. The below map was obtained from 501 papers 228 of which are about no tillage and 273 of which are about deep tillage

Adaptive measures for farmers

Improvement of CWP is widely accepted to be more complicated at plain scale than in field experiment as a result of interferences of factors such as crop cultivar, soil type, drought disaster, pest attacks, etc. There are good extension services in China to promote suitable technologies to farmers to increase. Deng et al. (2006) found that mulching with crop residues can improve CWP by 10%–20% by reducing soil evaporation and increasing plant transpiration. Straw mulching can be easily implemented and extended in the Huang-Huai-Hai Plain. Soil infertility is accepted to be another factor constraining productivity. The relatively low CWP across the plain suggests that achieving higher yield with the same amount of water consumption is possible for the plain and should be a goal for agricultural investments in water-scare region. However, the analysis of the results showed that this measure did not play a substantial role in improving CWP thus far. Consequently, adaptive measures such as cultivar alternatives, disaster and pest control, compensatory irrigation scheduling, residue mulching, etc shall be adapted for farmers to achieve CWP improvement with concentration of six agricultural sub-regions of the Huang-Huai-Hai Plain.

9.2.2 The uncertainties

Many studies have shown that drought, mainly agricultural drought, may become more severe and widespread under greenhouse gas (GHG)-induced global warming in the 21st century based on model projections (Burke et al., 2006; Cook et al., 2014; Sheffield and Wood, 2008; Zhao and Dai, 2016). There are still large uncertainties in recent and projected future drought conditions, especially regarding the extent to which the drought trends will be forced by changes in precipitation and evaporative demand (Sheffield et al., 2012; Taylor et al., 2013; Trenberth et al., 2014). Total drought events (based on SPEI-PM index) were described as no significant tendency over the historical period due to the drought duration, severity, and intensity reduced by the decreasing potential evapotranspiration. While, these results in our study are inconsistent with the previous studies, where Northern China was shown to have experienced a warm-drying trend (Wang et al., 2015a; Yu et al., 2014). This inconsistency is likely due to the use of different indices in previous studies and the variation in estimating potential evapotranspiration for different indices. Some studies have evaluated the use of PM equation to calculate drought indices and concluded that drought has changed little globally (Sheffield et al., 2012) and in China in the past decades (Wang et al., 2015b). Mcvicar et al. (2012) reported a global declining rate of evaporative demand of 1.31 mm·y^{-1} after reviewing papers on trends in estimated ET_0

($n = 26$), whereas Thomas (2000) found a decline rate of 2.3 mm·y^{-1} over China on the basis of a time series from 1954 to 1993. In our study, the annual averaged ET$_0$ across the entire Huang-Huai-Hai Plain revealed a pronounced decreasing trend of -1.29 mm·y^{-1}, which was higher than those over the Yellow River Basin (Liu et al., 2014b; Ma et al., 2012), the Yangtze River catchment (Gong et al., 2006; Xu et al., 2006), and the Haihe River Basin (Bo et al., 2011; Wang et al., 2010). These differences were mainly due to higher decrement of RH over the entire Huang-Huai-Hai Plain (as the governing climatic variable for ET$_0$).

Uncertainties in projections of climate change impacts on future crop yields derive from different sources in modeling. The relative impact of the change in each variable on wheat yield in isolation was conducted in this study using DSSAT-CERES-Wheat model under historical and RCP scenarios with the findings of positive impact of warming temperature and increasing precipitation on wheat yield in this study. Process-based crop models include the highest level of detail in simulating biophysical crop responses to multiple drivers of climate change and diverse farming management practices. However, these models need to be calibrated to a specific location and their aggregation for global scale climate impacts assessments, as done by Parry et al. (2004) and Nelson et al. (2009). Nonetheless, these crop models present some important uncertainties that need to be clearly identified and quantified as much as possible for robust impact assessments and sound decision making. First of all, some uncertainties in crop modeling results arise from uncertainties in the input data. Furthermore, unknown future socio-economic development and radiative concentration pathways (RCPs) necessitate comparison of different assumptions and scenarios of future GHG emissions, which directly infer with the climate system. In addition, uncertainties within soil and farming management data (e.g. crop calendar dataset, irrigated cropping areas and fertilizer application) required to drive crop models, and also within crop yields data used for model calibration and/or validation, will propagate during the process of crop simulations.

9.2.3 Referable value from dataset and methodology of this book

Some main findings put emphasis on investigating the extent to which the yield and crop water productivity for winter wheat in response to climatic warming and drought condition across the Huang-Huai-Hai Plain. Hence, the referable value is expected for other study and somewhere else from building-related dataset and developed methodology of this book.

The increased frequency of extreme climatic events has started breaking balances in hydrological cycle and resulted in large fluctuations in crop yields and water productivity in recent years. In addition, expected changes in soil moisture may alter the amount of water available to plant roots for transpiration (Goyal, 2004). Under elevated CO_2 conditions, the stomatal conductance in most species decreases, which may result in lower transpiration per unit leaf area (Kruijt et al., 2008). The potential impacts of climate change are expected to reshape the water demand and supply patterns, therefore it is essential to evaluate the impacts of climate change on water consumption and crop water productivity. Hence, the built dataset consisted of meteorological variables with daily time step under both historical 54 years and a high CO_2 emission scenario RCP8.5 can be applied to conduct the regional response of crop water productivity to climate change using process-based crop models.

And furthermore, this book documented an early plain-wide method integrated spatial and temporal assessment of CWP, ET_a, and grain yield with concentration of six agricultural sub-regions of Huang-Huai-Hai Plain. This study is the first attempt to apply regional evapotranspiration model and crop information to estimate actual evapotranspiration of wheat from double crop rotation system at a large regional scale. Besides, the county-level wheat yield map was further disaggregated to the pixel level using MODIS NDVI data during crucial growth stages as a bridge. The methodology can be easily replicated in other areas regardless of one single crop or one crop from double crop rotation system due to the simplicity of the process and the popularity of the data set required.

9.3 Conclusions

Drought and water shortage are generally accepted to be some of the most critical problems faced by the agricultural production of the Huang-Huai-Hai Plain. The overall ambition of this book was to investigate the extent to which the yield and crop water productivity for winter wheat could respond to climate change and drought using SPEI-PM method, DSSAT-CERES-Wheat model, SEBAL model and remote sensing data across the Huang-Huai-Hai Plain, which is the wheat production base of China.

Among our main findings, we highlighted the fact that: (1) an increase of ET_0 was predicted to lead to subsequent drought rise in frequency, duration, severity and intensity under the RCP8.5 scenario; (2) the cumulative probability of the simulated yield reduction was higher during jointing to heading stage in northern than in southern region due to water stress and changes in the management inputs; (3) the yield increase principally controlled increase of water productivity in north agricultural sub-regions and the increase of water productivity was more governed by increment of yield than the reduction of ET_a in other agricultural sub-regions.

Our results can have major agronomic consequences regarding the reform of the common agricultural policy in the Huang-Huai-Hai Plain, China. Despite insufficient precipitation for cultivation, Huang-Huai-Hai Plain, it is one of the main Chinese crop production centers, providing about 61% of the nation's wheat production. Furthermore, the farmers in the Huang-Huai-Hai Plain apply unplanned times, through flood irrigation during the winter wheat growing season to get high wheat yield. Consequently, Huang-Huai-Hai Plain faced the double challenges of maintaining high and stable crop yield and improving the crop water productivity of winter wheat by reducing water consumption. It will be adopted to develop feasible straw (film) mulching, regulated deficit irrigation, and soil water storage and preservation to reduce pressure on groundwater over-exploitation, especially for winter wheat in the Huang-Huai-Hai Plain. The results are expected to provide basic information for agricultural water management, improvement of crop water productivity and choice of adaptive mechanism under climate change in the Huang-Huai-Hai Plain. Finally, it is worth mentioning that our results have some uncertainties due to lack of data of deficit irrigation and fertilizer practices for CERES-Wheat calibration, and observed crop water productivity in agro-meteorological stations for temporal analysis, and the uncertainties arising from the input data of CERES-Wheat modeling. It is clear that further studies are needed to better understand how to implement these practices with emphasis of improving sustainability of these agro-ecosystems.

9.4 Prospects and improvements

9.4.1 Increasing RCP scenarios alternatives

In our study, only RCP8.5 scenario was selected to estimate SPEI-PM associated with ET_0 and the impact of climate shifts on wheat yield using the CERES-Wheat model. The RCP8.5 scenario describes the hypotheses about the highest population and relatively slow income growth with modest rates of technological change and energy intensity improvements (Riahi et al., 2011). A set of scenarios known as RCPs including RCP2.6, RCP4.5, RCP6.0 and RCP8.5 with different level of radiative forcing (Figure 9-3) has been adopted by related research (Moss et al., 2008, 2010). Projections of climate change impacts on drought and crop yields are accepted in our subsequent study with consideration of four RCP scenarios.

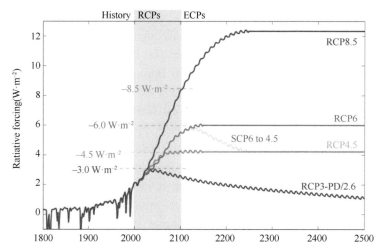

Figure 9-3 Total radiative forcing for RCPs, −supporting the original names of the four pathways as there is a close match on peaking, stabilization and 2100 levels for RCP2.6, RCP4.5 & RCP6, as well as RCP8.5, respectively (Meinshausen et al., 2011)

9.4.2 Increasing collection of irrigation and fertilizer management for DSSAT simulation

In this book, the potential impacts of drought on wheat yield were determined under a designed irrigation scheme from the estimation of precipitation deficit during growth stages at twelve stations representing different locations in the Huang-Huai-Hai Plain. Some uncertainties come from simulated results due to lack of consideration of deficit irrigation practices. Attia et al. (2016) found that a single irrigation of 100 mm at jointing or booting had increased grain yield by 35%, and 140 mm at anthesis or grain filling by 68% in the Texas High Plains. Zahe et al. (2014) observed that increases of the water and salinity stresses reduced seed yield and WUE and more usage of N fertilizer led to better WUE in West Azerbaijan, Iran. Chen et al. (2015) also observed that application of nitrogen and phosphorus fertilizers after the jointing stage increased the grain yield (112%) and WUE (96%) of winter wheat but resulted in a reduction of soil water storage by 12%, and that with the nitrogen and phosphorus fertilizers +plastic film treatment, the grain yield has increased by 53%, associated with WUE increase by 46%, and soil water storage increased by 21% after the jointing stage. Consequently, a long history irrigation station together with deficit irrigation and fertilizer practice and variety alternative is required for crop modeling in our study.

9.4.3 Increasing collection of observed CWP in agro-meteorological stations

The relationship between water productivity & ET_a and yield of winter wheat was conducted after partitioning the regional ET_a for wheat from the double crop rotation system at the level of the sub-agricultural zone in the Huang-Huai-Hai Plain. The ET_a was validated by latent heat flux estimated by SEBAL model and measured by eddy covariance during growing period of winter wheat during 2011–2012 in Yucheng station located in Northwest of the Huang-Huai-Hai Plain. As mentioned in previous chapters, the Huang-Huai-Hai Plain encompasses around 18 million hectares of farmland for wheat and maize double rotation production (He et al., 2009). More indispensable agro-meteorological stations need to be collected to validate the ET_a in our study. Although the crop water productivity of winter wheat was estimated for three periods of 2001–2002, 2005–2006 and 2011–2012, the combination with a temporal analysis of water productivity at agro-meteorological stations can help to reveal the whole picture of the relationships among CWP, ET, and yield in the Huang-Huai-Hai Plain in recent years.

9.5 Closing words

In the context of agricultural production found in the Huang-Huai-Hai Plain, our study was a pioneer in getting insights into water productivity at station and regional level. In Gembloux Agro-Bio Tech, before the start of the AgricultureIsLife project in 2013, for an over-exploitation of groundwater region where future warmer and drought conditions will heighten crop water demand, the questions about what the characteristics of ET and drought in various climate scenarios are and how winter wheat yield is affected by climate shift and drought were unclear. With our study, we have acknowledged that drought conditions will aggravate due to climate change by increasing ET_0 and augmenting ET_c, and that the relationship between water productivity & ET and yield of winter wheat can be defined at the level of the sub-agricultural zone in the Huang-Huai-Hai Plain. Finally, more indispensable agro-meteorological stations need to be collected to capture a comprehensive picture of the relationships among CWP, ET and yield. The results based on above prospects are expected to provide basic information for agricultural water management, improvements of crop water productivity and choice of adaptive mechanism under climate change in the Huang-Huai-Hai Plain.

Our study has led us to 'play' with various disciplines to answer our different questions from aspects of agronomy, microbial ecology, molecular biology, bioinformatics, and statistics. We point out that more holistic and multidisciplinary approaches are now required to improve our understanding of agro-ecosystem functioning with a view to achieving sustainability.

References

Ali M H, Talukder M S U. 2008. Increasing water productivity in crop production–A synthesis. Agricultural Water Management, 95(11): 1201-1213.

Allen R G, Pereira L S, Smith M, Raes D, Wright J L. 2005. FAO-56 dual crop coefficient method for estimating evaporation from soil and application extensions. Journal of Irrigation and Drainage Engineering-Asce, 131(1): 2-13.

Attia A, Rajan N, Xue Q, Nair S, Ibrahim A, Hays D. 2016. Application of DSSAT-CERES-Wheat model to simulate winter wheat response to irrigation management in the Texas High Plains. Agricultural Water Management, 165: 50-60.

Begueria S, Vicente-Serrano S M, Reig F, Latorre B. 2014. Standardized precipitation evapotranspiration index (SPEI) revisited: Parameter fitting, evapotranspiration models, tools, datasets and drought monitoring. International Journal of Climatology, 34: 3001-3023.

Bo T, Ling T, Kang S Z, Lu Z. 2011. Impacts of climate variability on reference evapotranspiration over 58 years in the Haihe river basin of north China. Agricultural Water Management, 98(10): 1660-1670.

Burke E J, Brown S J, Christidis N. 2006. Modeling the recent evolution of global drought and projections for the twenty-first century with the Hadley Centre climate model. Journal of Hydrometeorology, 7(5): 1113-1125.

Cai X L, Sharma B R. 2010. Integrating remote sensing, census and weather data for an assessment of rice yield, water consumption and water productivity in the Indo-Gangetic river basin. Agricultural Water Management, 97(2): 309-316.

Challinor A, Wheeler T. 2008. Crop yield reduction in the tropics under climate change: Processes and uncertainties. Agricultural and Forest Meteorology, 148(3): 343-356.

Cheema M, Immerzeel W, Bastiaanssen W. 2014. Spatial quantification of groundwater abstraction in the irrigated Indus basin. Groundwater, 52(1): 25-36.

Chen D, Gao G, Xu C, Guo J, Ren G. 2005. Comparison of the Thornthwaite method and pan data with the standard Penman-Monteith estimates of reference evapotranspiration in China. Climate Research, 28(2): 123-132.

Chen S, Zhang X, Liu M. 2002. Soil temperature and soil water dynamics in wheat field mulched with maize straw. Chin J Agrometeorol, 23(4): 34-37.

Chen S, Zhang X, Pei D, Sun H, Chen S. 2007. Effects of straw mulching on soil temperature, evaporation and yield of winter wheat: Field experiments on the North China Plain. Annals of Applied Biology, 150(3): 261-268.

Chen Y, Liu T, Tian X, Wang X, Chen H, Li M, Wang S, Wang Z. 2015. Improving winter wheat grain yield and water use efficiency through fertilization and mulch in the Loess Plateau. Agronomy Journal, 107(6): 2059-2068.

Condon L E, Maxwell R M. 2014. Feedbacks between managed irrigation and water availability: Diagnosing temporal and spatial patterns using an integrated hydrologic model. Water Resources Research, 50(3): 2600-2616.

Cook B I, Smerdon J E, Seager R, Coats S. 2014. Global warming and 21st century drying. Climate Dynamics, 43(9-10): 2607-2627.

Deng X, Shan L, Zhang H, Turner N C. 2006. Improving agricultural water use efficiency in arid and semiarid areas of China. Agricultural water management, 80(1-3): 23-40.

Dick W A, Mccoy E L, Edwards W M, Lal R. 1991. Continuous application of no-tillage to ohio soils. Agronomy Journal, 83(1): 65-73.

Döll P. 2002. Impact of climate change and variability on irrigation requirements: A global perspective. Climatic Change, 54(3): 269-293.

Esnault C, Stewart A, Gualdrini F, East P, Horswell S, Matthews N, Treisman R. 2014. Rho-actin signaling to the MRTF coactivators dominates the immediate transcriptional response to serum in fibroblasts. Genes & development, 28(9): 943-958.

Gong L, Xu C, Chen D, Halldin S, Chen Y D. 2006. Sensitivity of the Penman–Monteith reference evapotranspiration to key climatic variables in the Changjiang (Yangtze River) basin. Journal of Hydrology, 329(3-4): 620-629.

Goyal R. 2004. Sensitivity of evapotranspiration to global warming: A case study of arid zone of Rajasthan (India). Agricultural Water Management, 69(1): 1-11.

Guo R, Lin Z, Mo X, Yang C. 2010. Responses of crop yield and water use efficiency to climate change in the North China Plain. Agricultural Water Management, 97(8): 1185-1194.

He J, Wang Q J, Li H W, Liu L J, Gao H W. 2009. Effect of alternative tillage and residue cover on yield and water use efficiency in annual double cropping system in North China Plain. Soil and Tillage Research, 104(1): 198-205.

Immerzeel W W, Gaur A, Zwart S J. 2008. Integrating remote sensing and a process-based hydrological model to evaluate water use and productivity in a south Indian catchment. Agricultural Water Management, 95(1): 11-24.

Jiang J, Zhang Y Q. 2004. Soil-water balance and water use efficiency on irrigated farmland in the North China Plain. Journal of Soil Water Conservation, 18(3): 61-65.

Kruijt B, Witte J P M, Jacobs C M, Kroon T. 2008. Effects of rising atmospheric CO_2 on evapotranspiration and soil moisture: A practical approach for the Netherlands. Journal of Hydrology, 349(3-4): 257-267.

Lal R, Mahboubi A A, Fausey N R. 1994. Long-term tillage and rotation effects on properties of a central ohio soil. Soil Science Society of America Journal, 58(2): 37-46.

Li H, Gao H, Wu H, Li W, Wang X, He J. 2007. Effects of 15 years of conservation tillage on soil structure and productivity of wheat cultivation in Northern China. Australian Journal of Soil Research, 45(5): 344-350.

Li Y, Huang H, Ju H, Lin E, Xiong W, Han X, Wang H, Peng Z, Wang Y, Xu J. 2015. Assessing vulnerability and adaptive capacity to potential drought for winter-wheat under the RCP8.5 scenario in the Huang-Huai-Hai Plain. Agriculture, Ecosystems & Environment, 209: 125-131.

Liu E, Teclemariam S G, Yan C, Yu J, Gu R, Liu S, He W, Liu Q. 2014a. Long-term effects of no-tillage management practice on soil organic carbon and its fractions in the Northern China. Geoderma, 213(1): 379-384.

Liu Q, Mei X R, Yan C R, Ju H, Yang J Y. 2013. Dynamic variation of water deficit of winter wheat and its possible climatic factors in Northern China. Acta Ecologica Sinica, 33(20): 6643-6651.

Liu Q, Yan C, Yang J, Mei X R, Hao W, Ju H. 2015. Impacts of climate change on crop water requirements in Huang-Huai-Hai Plain, China. Climate Change and Agricultural Water Management in Developing Countries, 8: 48.

Liu Q, Yan C, Zhao C, Yang J, Zhen W. 2014b. Changes of daily potential evapotranspiration and analysis of its sensitivity coefficients to key climatic variables in Yellow River basin. Transactions of Chinese Society of Agricultural Engineering, 30(17): 157-166.

Ma X, Zhang M, Li Y, Wang S, Ma Q, Liu W. 2012. Decreasing potential evapotranspiration in the Huanghe River Watershed in climate warming during 1960–2010. Journal of Geographical Sciences, 22(6): 977-988.

Mcvicar T R, Roderick M L, Donohue R J, Li L T, Niel T G V, Thomas A, Grieser J, Jhajharia D, Himri Y, Mahowald N M. 2012. Global review and synthesis of trends in observed terrestrial near-surface wind speeds: Implications for evaporation. Journal of Hydrologys, 416-417: 182-205.

Mei X, Kang S, Yu Q, Huang Y, Zhong X, Gong D, Huo Z, Liu E. 2013. Pathways to synchronously improving crop productivity and field water use efficiency in the North China Plain. Scientia Agricultura Sinica, 46(6): 1149-1157.

Meinshausen M, Smith S J, Calvin K, Daniel J S, Kainuma M L T, Lamarque J, Matsumoto K, Montzka S A, Raper S C B, Riahi K. 2011. The RCP greenhouse gas concentrations and their extensions from 1765 to 2300. Climatic Change, 109(1-2): 213-241.

Molden D, Murray-Rust H, Sakthivadivel R, Makin I. 2003. A water-productivity framework for understanding and action. Water productivity in agriculture: Limits and opportunities for improvement. IWMI report, Sri Lanka.

Moss R H, Edmonds J A, Hibbard K A, Manning M R, Rose S K, van Vuuren D P, Carter T R, Emori S, Kainuma M, Kram T. 2010. The next generation of scenarios for climate change research and assessment. Nature, 463(7282): 747-756.

Moss R, Babiker W, Brinkman S, Calvo E, Carter T, Edmonds J, Elgizouli I, Emori S, Erda L, Hibbard K. 2008. Towards new scenarios for the analysis of emissions: Climate change, impacts and response strategies. Intergovernmental Panel on Climate Change Secretariat (IPCC) report, Geneva.

Nelson G C, Rosegrant M W, Koo J, Robertson R, Sulser T, Zhu T, Ringler C, Msangi S, Palazzo A, Batka M. 2009. Climate change: Impact on agriculture and costs of adaptation. Washington, DC: Intl Food Policy Res Inst.

Parry M L, Rosenzweig C, Iglesias A, Livermore M, Fischer G. 2004. Effects of climate change on global food production under SRES emissions and socio-economic scenarios. Global Environmental Change, 14(1): 53-67.

Pauw E D, Pauw W D. 2002. An agroecological exploration of the Arabian Peninsula. ICARDA report.

Piao S, Ciais P, Huang Y, Shen Z, Peng S, Li J, Zhou L, Liu H, Ma Y, Ding Y, Friedlingstein P, Liu C, Tan K, Yu Y, Zhang T, Fang J. 2010. The impacts of climate change on water resources and agriculture in China. Nature, 467(7311): 43-51.

Riahi K, Rao S, Krey V, Cho C, Chirkov V, Fischer G, Kindermann G, Nakicenovic N, Rafaj P. 2011. RCP8.5-A scenario of comparatively high greenhouse gas emissions. Climatic Change, 109(1-2): 33.

Ruud N, Harter T, Naugle A. 2004. Estimation of groundwater pumping as closure to the water balance of a semi-arid, irrigated agricultural basin. Journal of Hydrology, 297(1-4): 51-73.

Sheffield J, Wood E F, Roderick M L. 2012. Little change in global drought over the past 60 years. Nature, 491(7424): 435-438.

Sheffield J, Wood E F. 2008. Projected changes in drought occurrence under future global warming

from multi-model, multi-scenario, IPCC AR4 simulations. Climate Dynamics, 31(1): 79-105.
Tao F, Hayashi Y, Zhao Z, Sakamoto T, Yokozawa M. 2008a. Global warming, rice production, and water use in China: Developing a probabilistic assessment. Agricultural & Forest Meteorology, 148(1): 94-110.
Tao F, Yokozawa M, Liu J, Zhang Z. 2008b. Climate-crop yield relationships at provincial scales in China and the impacts of recent climate trends. Climate Research, 38(1): 83-94.
Tao F, Zhang Z, Xiao D, Zhang S, Rötter R P, Shi W, Liu Y, Wang M, Liu F, Zhang H. 2014. Responses of wheat growth and yield to climate change in different climate zones of China, 1981–2009. Agricultural and Forest Meteorology, 189-190: 91-104.
Taylor I, Burke E, McColl L, Falloon P, Harris G, McNeall D. 2013. The impact of climate mitigation on projections of future drought. Hydrology and Earth System Sciences, 17(6): 2339.
Thomas A. 2000. Spatial and temporal characteristics of potential evapotranspiration trends over China. International Journal of Climatology, 20(4): 381-396.
Trenberth K E, Dai A, Van Der Schrier G, Jones P D, Barichivich J, Briffa K R, Sheffield J. 2014. Global warming and changes in drought. Nature Climate Change, 4(1): 17-22.
Vicente-Serrano S M, Beguería S, Lorenzo-Lacruz J, Camarero J J, López-Moreno J I, Azorin-Molina C, Revuelto J, Morán-Tejeda E, Sanchez-Lorenzo A. 2012. Performance of drought indices for ecological, agricultural, and hydrological applications. Earth Interactions, 16(10): 1-27.
Wang H, Chen A, Wang Q, He B. 2015a. Drought dynamics and impacts on vegetation in China from 1982 to 2011. Ecological Engineering, 75: 303-307.
Wang M, Zhang C, Yao W, Wang X. 2001. Effects of drought stress in different development stages on wheat yield. Journal of Anhui Agricultural Sciences, 29: 605-607, 610.
Wang W, Peng S, Yang T, Shao Q, Xu J, Xing W. 2010. Spatial and temporal characteristics of reference evapotranspiration trends in the Haihe River basin, China. Journal of Hydrologic Engineering, 16(3): 239-252.
Wang W, Zhu Y, Xu R, Liu J. 2015b. Drought severity change in China during 1961–2012 indicated by SPI and SPEI. Natural Hazards, 75(3): 2437-2451.
Wright S F, Starr J L, Paltineanu I C. 1999. Changes in aggregate stability and concentration of glomalin during tillage management transition. Soil Science Society of America Journal, 63(6): 1825-1829.
Xiao D, Tao F. 2014. Contributions of cultivars, management and climate change to winter wheat yield in the North China Plain in the past three decades. European Journal of Agronomy, 52 (Part B): 112-122.
Xiao G, Zhang Q, Yao Y, Zhao H, Wang R, Bai H, Zhang F. 2008. Impact of recent climatic change on the yield of winter wheat at low and high altitudes in semi-arid northwestern China. Agriculture, Ecosystems & Environment, 127(1-2): 37-42.
Xin X G, Wu T W, Li J L, Wang Z Z, Li W P, Wu F H. 2013. How well does BCC_CSM1.1 reproduce the 20th century climate change over China? Atmospheric and Oceanic Science Letters, 6(1): 21-26.
Xiong W, Holman I, Lin E D, Conway D, Li Y, Wu W B. 2012. Untangling relative contributions of recent climate and CO_2 trends to national cereal production in China. Environmental Research Letters, 7(4): 044014.
Xu C, Gong L, Jiang T, Chen D, Singh V P. 2006. Analysis of spatial distribution and temporal trend of reference evapotranspiration and pan evaporation in Changjiang (Yangtze River) catchment. Journal of Hydrology, 327(1-2): 81-93.

Xu K, Yang D, Yang H, Li Z, Qin Y, Shen Y, 2015. Spatio-temporal variation of drought in China during 1961–2012: A climatic perspective. Journal of Hydrology, 526: 253-264.

Yu M, Li Q, Hayes M J, Svoboda M D, Heim R R. 2014. Are droughts becoming more frequent or severe in China based on the Standardized Precipitation Evapotranspiration Index: 1951–2010? International Journal of Climatology, 34(3): 545-558.

Zahe F K, Rad M N, Besharat S, Majnooni-Heris A, Jabbari A. 2014. Developing regression models between water use efficiency (WUE) and winter canola yield, under water and salinity stresses and different nitrogen fertilizer levels. International Journal of Basic Sciences and Applied Research, 3: 680-687.

Zhai J, Su B, Krysanova V, Vetter T, Gao C, Jiang T. 2010. Spatial variation and trends in PDSI and SPI indices and their relation to streamflow in 10 large regions of China. Journal of Climate, 23(2): 649-663.

Zhang J, Sun F, Xu J, Chen Y, Sang Y, Liu C. 2015. Dependence of trends in and sensitivity of drought over China (1961–2013) on potential evaporation model. Geophysical Research Letters, 43(1): 206-213.

Zhang J, Zhao Y, Wang C, Yang X. 2012. Impact simulation of drought disaster at different developmental stages on winter wheat grain-filling and yield. Chinese Journal of Eco-Agriculture, 20: 1158-1165.

Zhao T, Dai A. 2016. Uncertainties in historical changes and future projections of drought. Part II: Model-simulated historical and future drought changes. Climatic Change, 144(3): 535-548.

Zwart S J, Bastiaanssen W G, de Fraiture C, Molden D J. 2010. WATPRO: A remote sensing based model for mapping water productivity of wheat. Agricultural Water Management, 97(10): 1628-1636.

Zwart S J, Bastiaanssen W G. 2007. SEBAL for detecting spatial variation of water productivity and scope for improvement in eight irrigated wheat systems. Agricultural Water Management, 89(3): 287-296.